Organizing ITSM

Transitioning the IT Organization from Silos to Services with Practical Organizational Change

Randy A. Steinberg

 www.trafford.com
North America & international
toll-free: 1 888 232 4444 (USA & Canada)
fax: 812 355 4082

Other books by Randy A. Steinberg:

Implementing ITSM
Adapting Your Organization to the Coming Revolution in IT Service Management
Trafford Press ISBN: 978-1-4907-1958-0

Measuring ITSM
Measuring, Reporting and Modeling the IT Service Management
Metrics That Matter Most To IT Senior Executives
Trafford Press ISBN: 978-1-4907-1945-0

Servicing ITSM
A Handbook of IT Services for Service Managers and IT Support Practitioners
Trafford Press ISBN: 978-1-4907-1956-6

Architecting ITSM
A Reference for Architecting and Building the Entire IT
Service Management Infrastructure End To End
Trafford Press ISBN: 978-1-4907-1957-3

An IT executive said,

Organizational Change here is dead.

Great IT technicians are we,

Who can work separately?

Let the Service Desk do it instead!

It's time to operate IT like a Service Organization.

- The Author

Dedication

This book is dedicated to those very hard working IT professionals, managers and executives who deserve to see their IT solutions deploy and operate day-to-day within acceptable levels of costs and risks to their company.

Table of Contents

Chapter 1

Book Overview

Hey…Is That A Million Parts Flying In Formation?
…Or Just An Airplane?

Organizational Change: a process in which an organization changes its working methods in order to develop and deal with new situations or deliver new kinds of services.

Transitioning an IT organization from traditional technology silos to an effective IT Service Management (ITSM) delivery organization is not easy. While implementation of new processes and tools are generally understood by IT teams, many of those teams struggle with the organizational aspects of their efforts. New software and hardware may get successfully implemented, but efforts fall down on communication issues, lack of training and reluctance to doing things differently by those who are to use the new system.

This book will directly address the activities, steps and approach for executing on a program of organizational change to overcome those challenges. It is written specifically for the ITSM practitioner working on ITSM initiatives. Rather than dwell on organizational change theory, it provides a practical approach, used successfully many times on ITSM projects, for addressing the "people" part of their efforts in order to be to be successful.

Why This Book Was Written

IT is rapidly moving from a focus on engineering capabilities to a focus on integration of services from many sources. Advances in cloud computing, virtualization of physical IT resources, outsourcing, hosting, co-location and many third party IT solutions are offering more choices than ever to meet new business needs offering lower cost points and faster delivery times. What once had to be internally engineered can now be bought in the marketplace like pieces and parts of a jig saw puzzle. Restructuring and new working methods are needed to pull these pieces together for the business. Continuing to operate in technology silos without a service focus will only spell more and more trouble for IT organizations.

1

Moving from silos to services cannot be done without a serious effort in helping IT organization staff make this transition. You cannot simply design new processes, implement new tools and imbed a customer focused service culture and expect that people will make it happen. They must be carried along the way through a program of organizational change or the initiative will fail.

The practice of organizational change is little known within IT organizations. It's not taught in school, there is no training in the discipline for IT workers and the skill sets involved are quite different than what most IT organizations are prepared for. While the disciplines of organizational change have been around for decades, very little content appears geared to the IT organization.

Much of the literature and practices around this tend to be of a theoretical nature. IT organizations find it hard to link those concepts to how they should actually transition people to accept, support and operate IT solutions and services. There may be temptation to just skip the change effort altogether or have senior executive management decree "just do it or go elsewhere". Both of those strategies fail. Organizational change issues can make or break an ITSM initiative. Forcing people to take on the change without winning hearts and minds typically fails within months.

This book is written to specifically overcome the gap between organizational change theory and what an ITSM program needs to do. Guidance and practices shown here are taken from actual IT and IT Service Management implementation efforts. It is written from the IT Service Management perspective with specific practical approaches that have been used successfully in other IT organizations. It is hoped that the content in this book can serve as a reference guide to IT workers, be they executives, middle management or project leads who are working on service management initiatives to help make them successful.

IT workers hate anything to do with organizational change. In their view, it's political, soft, not technical and a big waste of time. Many just want to get the implementation work done and move on in the mistaken belief that good people will learn new ways of working on their own. Yet consider some of these real situations that actually happened:

> A major credit card company whose senior executive management was completely onboard to transition to IT Service Management which subsequently failed because IT support staff didn't want anything to do with it.

> A major consulting firm that did a very well done state of the art manufacturing system for a major steel making company that ended up getting sued because "the system was too hard to use" and therefore useless to the company.

> A major bank IT organization that spent millions investing in new IT Service Management tools and processes only to see that never roll out and the implementation team isolated from the rest of the IT organization ("…it was their project, not ours…we didn't really do things that way…").

Properly executed organizational change management can make the difference between success and failure for any IT implementation project including ITSM. It is critical for any IT Service Management transition. Yet how to accomplish this?

IT is comfortable with technology. Yet when it comes to organizing people so that they understand the value of IT Service Management, adhere to schedules, and ultimately adopt the solutions being implemented, this can seem like some sort of "fluff" that takes away time and resource that could be better spent on just getting the hardware and software implemented.

The good news is that organizational change management is far from rocket science. The approach and techniques discussed in this book can be easily understood. If you apply a good mix of tools from the change management bag of tricks –– and many are shown as part of this book - there's no reason you shouldn't succeed.

The IT Organizational Change Challenge

IT has historically grown up in vertical technology silos. This can no longer work in an age where many IT solutions can come from many outside sources. Someone needs to be accountable for the actually delivered business support services that depend on all those pieces and parts. Without this, communication and integration across silos are forced upon IT executives – a role they have neither time nor desire to play. If executives avoid this, then integration will fall to the end customer with even worse consequences.

Too often executives make the mistake of hoping that better processes and practices alone will make everyone play nice together and cooperate across the silos. This seldom works by itself. Today's IT organization must also be organized to deliver horizontally across vertical technology silos or they will soon become ineffective. The horizontal part is what we refer to as "services". Their value is create a seamless experience for customers such that end users and customers do not have to be bothered by the specifics of the underlying IT technologies and infrastructure.

Some executives may be tempted to go down the "dictator" path to avoid spending the time and trouble on organizational change. After all, get rid of a few dissenters and the staff will get the message. I've personally seen this approach tried a number of times in IT organizations. It works for about the first few months and then soon collapses on itself either because the resistance among support staff gets stronger or the executive leaves the company or moves elsewhere. I have never seen it work. The people effect needs to be dealt with.

When IT organizations are asked to list their top barriers and top success factors in getting things done, they listed these items:

Top Barriers	Top Success Factors
Functional Boundaries	Executive Sponsorship
Lack Of Change Skills	Treating Staff Fairly
Lack Of Communication	Involving Employees
Employee Opposition	Frequent Communications
Lack Of Training	Sufficient Training
Initiative Fatigue	Focus On Cultural Skills
Middle Management	Using Internal Champions

Notice that both barriers and success factors have people issues at their core. Whether you are implementing IT Service Management or any other IT solution, you are changing the way people work and operate. Efforts are needed to address that change or people will go down a wrong track and cite your efforts as having failed.

IT support staff also present some challenges. Historically, they will get caught up in implementing technologies or addressing failures. This may make them too busy to spend time on organizational change activities. In reaction to daily incidents and problems, they will cast organizational change efforts aside under pressure to focus on delivering those services safely. IT projects under tight deadlines may skip organizational change tasks because they were not planned for and perceive that they will only delay project completion.

IT support and operations staff, especially like things cookie cutter, consistent, and not to change. They are extremely change averse and many times for good reason. Deviation creates mistakes, delays and outages. Only under pressure from a significant event like a major outage, bad publicity, and major customer loss will they look for a change or view things differently.

The biggest challenge of all? You can't take an IT organization down to retool. Organizational change activities must take place in parallel with current operating and support practices. Change must be implemented while maintaining organizational and operating continuity.

Book Chapters in Brief

Brief descriptions of remaining book chapters are as follows:

Chapter 2 – Overview of Change

This chapter presents an overview of what Organizational Change is really about. It addresses the fundamentals of human behavior and how to adapt that behavior to new ways of working. It also describes the infrastructure and eco-systems that need to be in place for any successful change program.

Chapter 3 – The Service Driven Organization

How should IT organize as a service provider? This chapter tells you how. It presents the key service roles that need to be in place for IT organizations. It also presents what is in place at many successful IT organizations who are transforming to IT Service Management.

Chapter 4 – ITSM Organization Models

This chapter covers the different types of organizational models used to operate ITSM. Pros and Cons are presented with each model as there is no one correct organizational answer for all IT organizations.

Chapter 5 – Building The IT Organization

This chapter covers means and methods for how you might build and design an IT organization. It includes a discussion on the RACI approach for defining responsibilities as well as a transformational approach for ITSM organizations.

Chapter 6 – Dealing With Resistance

This chapter discusses why resistance exists and what you can do about it. It identifies the different types of people who might put up roadblocks and gives you suggested approaches for dealing with each one. It also provides some warning signs that can act as early warnings that change may not be taking hold.

Chapter 7– Building Communications Step By Step

The meat of this entire book, this chapter introduces you to an organizational change program approach designed specifically for IT organizations. It describes how the program would be put together, how the change team should be resourced and provides an overview of the key work stages for the program. This approach has been used successfully with IT organizations undergoing an ITSM transformation.

Chapters 8-12 – Communications Work Stages

These chapters discuss the communications change program in detail. Each chapter addresses one stage in the approach identifying the key activities and outputs for that stage. Additional guidance is provided around techniques and approaches that can be employed during each stage.

Chapter 13 – Communication Tools and Techniques

This chapter highlights some of the many tools and techniques that can be used as part of your change program. It includes examples such as Kaizen, Brainstorming, Information Mapping and many others. Other sources are also identified where you can get additional guidance beyond what is in this book.

Chapter 14 – ITSM Operating Roles

This chapter presents the key operating roles for an ITSM program along with roles descriptions and recommended skill sets.

Chapter 15 – Training Considerations

This chapter presents various strategies that can be employed for training, training production, and training delivery. It also presents an approach from a user point of view covering various stages of learning from awareness to mastery.

ITSMLib Download Site

There are a number of tools that you may find helpful related to this book. These can be downloaded through a facility called ITSMLib™. This facility provides access to real world working documentation, templates, tools and examples for almost any ITSM project. The library is structured to easily find knowledge and can be easily searched with phrases and keywords to find relevant information. You may find this useful in jump starting your own solutions with ideas and content that has worked for others.

Included in this site is an entire sub-library titled Organizational Change. Here you can find many tools and aids that can help you jump start your organizational change efforts. Requests for access can be made to:

RandyASteinberg@gmail.com

When requesting access, please provide:

1) Your Name

2) An Email address

3) The company you work for

4) The country you reside in

This library is continually updated with content and we try to add things all the time as people see a need.

Chapter
2

Overview Of Change

The Organizational Change Mission

What is the bottom line of what we are really trying to accomplish?

ITSM solutions almost always involve changes in the way people perceive their roles, how they operate, and what methods and tools they use. The collection of new tools, roles and processes represent a new culture in the way things will work. With ITSM, we are replacing that existing culture (traditionally rooted in technical silos) with a new culture (service-based delivery organization).

At the same time, the existing culture is already embedded into how people think within that culture. That thinking is "cemented in" with current hopes, behaviors, attitudes, beliefs and other aspects that influence how they make decisions in the workplace. If we attempt to operate with a new culture, but leave people with their existing thoughts and beliefs, we will face tremendous resistance.

Therefore, the goal of ITSM organizational change activities should be to not only communicate a new culture, but to help people break away from the old behaviors that now conflict with that new culture. A program of organizational change needs to be in place to transition and embed new behaviors into the existing culture, transforming it to the new culture.

Consider this approach (which happens all too often):

A small ITSM team goes behind closed doors and designs new Incident and Change processes. After a month of work, the team documents and publishes those new processes. They then communicate how the new processes work and tell the rest of the IT organization that they will have to start using them Monday morning.

What do you think then happens?

Most likely, this approach will result in a lot of resistance. IT support staff not involved in designing the new processes will feel they have been forced upon them. They may cite that the process is not complete and won't work in their situation. Mounting frustration will cause IT management to delay the rollout of the new processes. This in turn will send a broad message that the process solution is no good, not usable and won't work. Politely, they will delay the rollout citing other priorities. The process team will become isolated. "Good work", they will say, "but doesn't really apply to the way the rest of us do things".

The diagram below presents a summary of what we are trying to accomplish with organizational change:

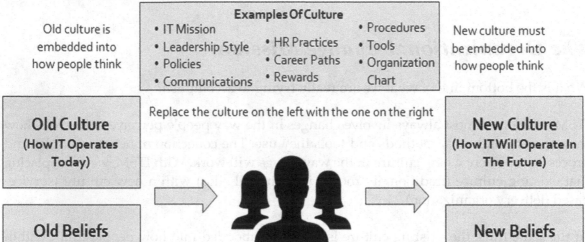

There are two key things we are trying to do:

1) Communicate the new culture so people will understand how things will be done

2) Undertake activities that will help people transition their thoughts and beliefs towards that new culture.

Too often we focus on the first task, but ignore the second. This is where the line is drawn between a successful ITSM transition and a failed one (or one that might be moving much slower). You can't simply force people into the new culture. Their hearts and minds must also accept the changes that are being imposed on them.

Current IT Organization Challenges

Most IT organization structures are falling out of date with the shifts created by the introduction of new technologies, expanding services, pace of business change and the overall trend of lifting IT development and support activities up and out of the traditional data center. Many activities that required internal engineering can now be bought on the open market and from many sources.

Remember the days when applications were developed on a few carefully chosen selected platforms? In today's world, application solutions are now split into myriads of pieces and parts that have many dependencies. One example might be a .NET written application that is dependent on a specific set of browsers, writes to a database provided by another vendor, utilizes CSS and HTML for presentation code, includes a host of separate graphics files for displays, provides permissions through a separate set of directory services, and makes API calls to a 3rd party vendor product.

Now add in that the application solution now sits on a set of virtualized servers located at a 3rd party hosting site, sits behind a series of customized firewalls, accesses storage devices through a local storage area network and has its client end sitting on virtualized PC devices. We won't even mention the remote tablets and phone "apps" that are also being used.

Get the idea? Can an organization that is solely organized by technology function even function in this environment? Should we continue to add more technology organization silos for each new piece of technology? Who puts the pieces together? Are we leaving it up to executive management to do this, or even worse, the end users and customers?

This is why IT Service Management is so important. The details of integrating, managing and supporting all this stuff needs to be hidden away from end users and customers. The means for hiding it is through IT services. The IT organization needs to be organized as a service provider, not an engineering company. The need for more effective ways to communicate across the organization and provide accountability for what is delivered to end users and customers is more critical than ever. Yet, these items are missing in most IT organizations.

The ITSM Organizational Models chapter provides some guidance for selecting an organization approach that may best work for your organization. While there is no one right answer, this chapter provides you with a good overview of common IT organization models and lists the pros and cons with each one as it pertains to running a service delivery organization.

Guiding Principles For Change

When asking ITSM project leaders which tasks they spend the most time on, here is what they commonly cite:

- Communications with stakeholders

- Getting stakeholders on board

- Managing conflicts

- Getting teams to collaborate

- Developing team skills

The combination of the above tasks was cited to take up 70-80% of the ITSM implementation effort. In other words, communicating and managing conflicts with changes to ITSM practices rises to the top in the list of challenges for these kinds of efforts.

Effective ITSM solutions will require working across traditional organizational boundaries to engage the whole enterprise. Organizations must address with both cultural and structural barriers in a way that involves the whole enterprise if they are going to be successful in the complex world they operate in.

So what seems to be working?

Listed here are some general guidelines to consider as you undergo an ITSM effort. From experience, with many IT organizations it boils down to:

Principal #1: All change should be tightly linked to a compelling shared purpose, driven by strategy and a way to measure success.

Purpose communicates why the change needs to happen. It tells people that what is happening is important to position the organization for success in the future in some way. Without this, the ITSM initiative will be seen as something that will eventually go away and die on the vine.

The purpose must have three things that need to be present:

- Dissatisfaction with the current state that is public and shared. People must be dissatisfied with the status quo and feel some action must be taken. The dissatisfaction needs to be present, but not at levels so severe that people are giving up, see no hope in changing anything and looking elsewhere to work.

- There must be a compelling and ennobling picture of the future that is also public and shared. This vision must seem real enough that people can see a path from where they are today to where they will be tomorrow.

- Immediate steps that can be taken today towards that vision that will produce immediate evidence of progress.

If any of the above is missing or poorly communicated, there will not be enough momentum generated to overcome the resistance. If all of the items are strongly communicated, then change will be successful.

Principal #2: Use a holistic strategy to drive the change forward that considers all aspects of what needs to be done to drive the change forward.

The key aspects of the change strategy to consider are:

- *Strategic Direction:* Strategy will drive everything. The change must have purpose and direction. It includes a clear statement of mission, vision, values, goals and objectives.

- *Systems:* This refers to the systems and processes used to accomplish the work required to effect the strategy. Are we doing the right work? Is it being done in the best way? Do we have the right systems (e.g., information, rewards, hiring, training systems) to support the change?

- *Organization:* Organization charts and job descriptions are part of this, but so is the way decisions are made and the informal relationships that are used to get the change done. The distribution and use of power resides here. Are we organized correctly to effect change?

- *Resources:* Resources are what is needed to achieve the strategy. This includes motivated people who understand and are committed to the strategy. It also includes people who have the skills, equipment, facilities and technology to get the work done that will achieve the strategy.

- *Leadership:* Aligning and building leadership for change.

- *Stakeholder Management:* Engaging people to accelerate change. This involves identifying who the key stakeholders are and their role in effecting the change. It also includes lots of communications as transition towards the change take place. Momentum will be highly dependent on how stakeholders engage and support the strategy and vision.

- *Roadmap:* Communicating a shared change journey that is visible to all who are taking it that reflects and integrates all change milestones and initiatives.

Principal #3: Share information that lets everyone have the big-picture view they need to act promptly and effectively to support the strategy.

This can't be communicated enough. Don't just issue communications once and assume that everyone has understood the message. Keep communicating information over and over throughout the entire change effort. Communication to stakeholders needs to cover:

- *Vision:* What is our future picture of success? How would we describe, in detail, the world that we are seeking to create where we are operating successfully?

- *Mission:* Why do we exist? What is our fundamental purpose for ITSM as well as our IT organization? What "business" are we in? What "services" will we provide? How do we create value for our business stakeholders?

- *Environment:* What is happening in the world that we need to take account of in our planning? What are the forces, trends, technical and business developments that impact our ability to succeed?

- *Stakeholders:* Who has a stake in our organization? Who is counting on us for something? This can include customers, employees, owners, suppliers, communities, unions and others.

- *Values:* What are the guiding principles that we stick to even when the going gets tough? What are the behaviors that really define how we will operate in the new world?

Principal #4: Measure the right things and communicate the results as close to real time as possible.

What are the measurable, attainable outcomes to be achieved in the next year? How will we know we are making progress on our goals in pursuit of our vision? For ITSM initiatives this can include things like:

- Fewer incidents and outages

- Lower operating costs

- Lower business costs

- Faster capability to make changes to meet upcoming business needs

- Free the business to enter new markets and customers

- Improve customer or business satisfaction with IT services

- Increase revenues

- Decrease penalties

- Shift labor from reactive tasks to proactive improvements and developments

- Raise IT workforce morale

- Effect a consolidation, merger or acquisition

- Deliver IT services faster and at lower cost points

Principal #5: Effect the ITSM change with a clear overall operating plan, roles and responsibilities.

Don't just implement new ITSM tools. Develop comprehensive plans that address processes, data, roles and responsibilities, governance and how technology will integrate and support all of these. Organizational change itself should also be an additional track of effort that runs parallel throughout the entire plan. More detail on what such a plane should look like is covered somewhat in this book and in depth in the Implementing ITSM book.

The plan itself should include:

- The tasks to be done
- Who is accountable for doing the tasks
- Deliverables or outcomes from the tasks
- Key milestones
- Key timeframes

A master plan should be developed, and then shared across all teams and stakeholders. Progress against this plan should be clearly communicated throughout the entire change effort.

Principal #6: Consider the impacts of changes in stakeholders.

Stakeholders are not permanent. They will come in and out of IT projects. They get promoted (or demoted) and move on to other things. Considerations for this should be considered as part of the change planning effort. Items to watch out for include:

- Loss of momentum for the change due to changes in leadership. New leaders may not have been part of communications beforehand and may come in with other agendas. Special care needs to be taken to communicate with new leaders as soon as possible and frequently.
- Delays in ITSM progress caused by changes in support staff. New staff also need to be brought up to speed as soon as possible.
- Disruption in momentum caused by organizational changes, promotions or shifts in organization responsibilities. For these, have a strategy that regroups everyone together to assess the impact on current change activities and what can be done to continue those activities with the new organization.

Principal #7: For each member of the leadership team, it must be clear that change is the work and not a distraction from the work.

Leading a significant organizational change effort requires a shift in thinking for many IT executives. The change effort requires a significant time commitment that they are often not prepared for. Often 50-60% of a leader's time is spent working on issues related to the change.

In building the Leadership Team, the focus is on building a sense of membership around the strategy. If everyone on the team feels heard and commits to the strategic direction, the team will have a firm foundation for working. The team will also need to work out its own norms and mechanics for operating. To successfully lead strategic change, a Leadership Team must also be aligned around

- Membership - who is on the team

- Roles – what is each leader's role in leading change

- Control- how decisions will be made and by whom

- Goals - what results must be achieved

Principal #8: Establish a program of organizational change with the primary goal of building consensus around the changes that need to take place.

The key effectiveness of the Organizational Change approach is the idea of engaging a critical mass of the organization in the change effort. This cannot happen by itself or with good intentions. It must be planned for and operated as a critical part of any ITSM effort. Some examples of techniques to obtain positive engagement include:

- Getting people involved in helping to define the solution. Almost no one resists a solution where they feel they had a part in building it. Find ways to get people involved and let them have their say.

- Communicate often and communicate frequently. Plan out a series of communication events that build over time to increase momentum. Do not assume that just because people hear things once, they fully understand and agree with the changes taking place.

- Communicate success stories and progress as the change effort executes. Don't limit communications to project teams and team leadership. Make sure everyone sees a growing success story that they can become part of.

- Make sure senior management is engaged. They must be seen as leading the effort. This can include having them participate in key meetings (even if only briefly), quoting them in company newsletters and announcements and mentioning the change effort at key company events.

- Communicate a clear roadmap for change and what people can expect along the way. Let people know when key milestones will happen and how their work approach may change over time.

While these principles may seem a bit generic in nature, they underlie many of the techniques and approaches discussed in this book.

Creating The Infrastructure For Successful Change

When planning and implementing an ITSM initiative, the infrastructure for successful change must be put into place. This infrastructure consists of organization, communication, measurement and human resource systems that promote and enforce the changes about to take place. Here is some suggested guidance on what to put into place:

Team Organization

Utilize a Core/Extended/Advisor team approach as a means for effecting change across the IT and business organization. This approach appears to work well where change needs to take place in medium to large IT organizations. A brief description:

Core Team – (Typically Full Time)

This represents project team members who are doing heads down design, build and implementation of ITSM solutions. Members of this team are usually assigned full-time throughout the project. They are the hands-on people that will get ITSM solutions built and implemented.

Extended Team – (Typically up to 4 Hours Per Week)

The workhorse of the organizational change effort. This team consists of key business and IT stakeholders from all the business units in the enterprise who will be impacted by ITSM activities. They have two major responsibilities. First, they work closely with the Core Team to provide feedback on process, technical or organizational design decisions. Secondly, they represent the buy-in agreement for the business unit (or units) they represent.

This means that they carry key ITSM Program strategies back to their business unit, shop them around and obtain agreement within their unit. Alternatively, if all is not okay, they bring back concerns and issues to the Core Team. This greatly leverages the Core Team and allows the Core Team to focus on design and build activities versus getting sidetracked in efforts to go to each business unit in the enterprise to review what is being done. Members of this team usually are assigned part-time, but may have periods of time where more or less commitment is needed.

Advisor Team – (Typically up to 1-2 hours per month)

This represents senior management personnel with a vested stake in the ITSM solution. Members from this team are used to provide final decisions on behalf of the business units they represent. They may be called on to assign an Extended or Core Team member to assist with the change effort. While typically 1-2 hours per month is a good norm, experience has shown that some executives may simply desire to be notified of key implementation activities and events when they occur. Others may want to get more fully involved.

These teams work together to fan out change across the enterprise. The Core Team builds the solutions, the Extended Team pushes the solutions out to the rest of the Enterprise and the Advisor Team steers the solution and handles any issues that require senior leadership attention.

Event Planning Team

The organizational change effort will essentially consist of a series of events designed to communicate the changes about to take place. Therefore, there is a need to plan events, schedule them, register participants, track who attended events and coordinate the event itself. Resources should be established to conduct these activities. Without this team, ITSM project resources will get bogged down in organizing and scheduling these activities taking time away from developing their solutions.

Measurement Systems

What gets measured, gets done and in many cases, some very important things may end up unmeasured. Systems need to be in place to track and measure activities and the results they achieve. Measurement is feedback and that feedback is only useful if it is relevant, current and actionable. Some examples of measurements can include:

- Adherence to implementation schedules and timelines

- Progress towards attainment of new skills and capabilities

- Attendance at key communication events

- Progress on key outcomes to be achieved by ITSM implementation activities

- Certifications for new skills and technologies

- Timely resolution of program issues and problems

Communication Systems

Information flow is critical in times of change. In the absence of facts, gossip will flourish. Communication of what is happening and what is being accomplished can be done in high tech and low-tech ways. Web sites, Intranet, workplace kiosks and video conferencing are all good. Low-tech channels such as bulletin boards, newsletters, banners and memos dropped on a desk can also deliver information effectively. Town Hall meetings, brown bag lunch briefings and one-on-one conversations serve the added function of allowing information to be collected as well as disseminated.

To support email communications, up front activities will be needed to establish distribution lists and methods for building, vetting and approving distribution of communication emails. There is nothing worse than having to deal with misleading emails, emails sent to the wrong parties or emails that send out conflicting messages to stakeholders.

Rewards and Recognition Systems

Measurement systems must be effectively linked to rewards and recognition. The annual goals of managers, for instance, need to be linked to measures related to achieving ITSM changes. Some rewards may need to be based on team rather than individual performance. Rewards don't have to be monetary. Surprising people with praise or a reward is a great motivator. Sometimes a simple recognition during a key meeting or company announcement is the most valuable reward you can give. What is important is that the

Rewards and Recognition systems are used to demonstrate progress towards change that can be seen by others.

Training Systems

For acquisition of new skills, training systems need to be in place to procure, register, schedule, track and report on training events and activities. If many stakeholders are involved, this can become a full time effort that needs to be carefully managed. See the chapter on Training Considerations in this book for much more detail.

Human Resources

Systems need to be in place for hiring new staff, changing staff job roles and linking job performance to rewards and recognition. The ITSM solution may require recruiting efforts to build support staff. Some staff may be transferred out or in depending on the ITSM changes taking place. There may be changes in pay and benefits as a result of the changes taking place.

Financial Support Systems

Systems should be in place to support budgeting, accounting and auditing of change activities. Purchase Orders may need to be cut to obtain new hardware and software. Billing and payments need to be made to vendors and consultants that may be involved with the changes taking place.

Procurement and Legal Systems

These would be needed to negotiate large purchases or contract with solution providers. Many IT organizations may have strict requirements around how things go out for bid or ensuring selected providers and products follow organization policies for ensuring fairness and dealing with social and union requirements.

The Service-Driven IT Organization

Organizing As An IT Service Provider

Most IT organizations are organized around technology silos. Business users consume "services" which are underpinned by one or more of these silos. In an IT organization structured by technologies, the accountability for the bundling of those technologies into what business users actually use is missing. This leaves the integration of those technologies at the senior management or executive level. If not done at that level, then business users end up getting involved with all the IT details.

Putting executives in charge of improving cross-functional processes in large organizations is akin to asking them to paddle a boat upstream. The natural forces in the organization will tend to frustrate their every attempt to coordinate activities across functional boundaries. A symptom of this is a major incident that crosses multiple technology silos. Executives find themselves on the phone having to call multiple technology teams to determine a course of action. The natural reaction of each technology team called is to focus more on why their team is not responsible than coordinating to fix the issue.

So what does it really mean to be a service organization? Why should IT care about services? What's the purpose of having services anyway? What do services mean to an IT organization?

The answer to these questions is that the typical IT organization is utilizing an engineering-based delivery structure just as it has for decades even though the world IT now lives in has been rapidly changing. Much of what IT used to support, build and deliver can be now be purchased elsewhere. IT is being lifted up and out of the IT data center with Cloud-based "anything-as-a-service" offerings. New technologies and devices are spreading and stretching when and where IT services can be used (e.g. Internet Of Things). A plethora of DevOps practices and tools are accelerating delivery of IT solutions, combing development applications with supporting infrastructures. Today's IT solutions are no longer single application programs, but now made up of hundreds if not thousands of independent parts. Best practices in capacity management now forecast and grow through containers, stacks and pods of combined technologies versus the old way of forecasting by individual technologies.

Yet across all of this, IT is still mostly organized into discreet technology units with little or no accountability for what the business and end users actually see. As a result IT executives and customers end up working with the many disparate IT units to get things done, with tremendous frustration.

So here is the proposition for why services are so critical to IT. Something needs to shield the business from the complexities that are inherent in today's world of IT solutions. A "service buffer" needs to be in place. Accountability for each service needs to be in place. IT career paths need to diverge allowing for technology specialists who understand how things work on an equal footing with those who understand how to bundle and integrate disparate technologies and service offerings from others into a seamless service that customers can understand and consume.

The model below shows how this might fit into today's traditional IT organization:

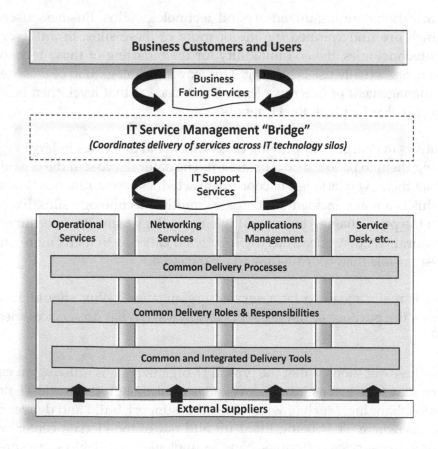

The above model hides the complexities of activities done by the traditional IT engineering silos behind services. Each service is owned "end-to-end" by a Service Owner that cuts across traditional functional lines. The Service Owner is responsible for delivery of the service across organizational boundaries (combination of line and cross function reporting).

Other related roles include:

- **Service Manager** – handles day-to-day delivery of the service

- **Process Owner** – owns and manages one or more processes used to deliver the service

- **Business Relationship Manager** – owns relationships with customers and end users who receive the service

- **Technology Owner** – owns and manages one or more technologies used to deliver the service.

These all work together towards providing seamless delivery to the customer. These roles are described in much more detail later in this chapter.

The Service Owner Is Everything

The first tendency by executives is to try to get the silos to work together. After all, they're led by good smart people who should know how to work together. In reality, they will be focused on the missions of their individual silos, priorities and objectives. Even if they try to work with other teams, they don't have clear organizational support to ask those teams to change priorities and efforts.

The role of Service Owner is the most important organizational change needed to overcome this deficiency. With this role, a Service Owner is accountable end-to-end for delivery of the service. If the service experiences an incident, it doesn't matter where the root cause lies or which team needs to be called in. The service is still out and it is a Service Owner responsibility to get the service back up, even if that includes coordinating with many teams to make that happen. No more pushing tickets back to the Service Desk!

It is strongly recommended that a Service Owner role be established and at an organization level that allows them to interoperate across existing technology organization silos. This role should be placed within a new organization silo that reports directly to the IT executive leadership. As an alternative, this role could exist within a logical organization structure drawn from the existing silos. In this case, the logical organization has a direct line to the IT executive leadership. See the chapter on IT organization models for some further thoughts.

Make the role permanent and incorporate it into the overall IT organization. They not only provide a single point of contact function for a service, but also work to improve service delivery performance. Broad-based measures that cut across technology silos are used to gauge service performance. The failing of one or more technology silos no matter where they exist will be reflected in the incident counts, problem counts, service breaches and delivery measures for the service as a whole. Incorporate these measures into a scorecard or dashboard for the service.

Select service owners with strong leadership skills and develop those skills even further. Choose owners who not only have experience in delivery of those services, but also have charisma, good relations with senior leaders, political savvy, and the ability to persuade. Then explain the role, set expectations, and further train and coach them in service improvement, persuasion, and coaching.

Make the service owner accountable for how well the service performs. Include in their job description the responsibility accountability for taking action when service performance is out of spec or trending out of spec.

Give the service owner organizational power. For example, they should have access to senior leaders, involvement in all major and monthly review meetings, budget control, a title, and a say in rewards and incentives for line managers.

To ensure that a service is managed with a business focus, the definition of a single point of accountability is absolutely essential to provide the level of attention and focus required for its delivery.

Habits Of Effective IT Service Organizations

Whether you are responsible for supporting distributed systems, internets, or traditional legacy systems you have most likely had more than your share of fire-fighting, emergencies and run-ins with end-users, applications developers and company management who appear to have little appreciation for your efforts. Take some time out and review this list of common habits observed in other support organizations who have somehow seemed to find a path to success.

So what is working inside those IT organizations that are delivering good service? What characteristics do they share?

Service Accountability

Service Owners are in place and held accountable for the quality of the services they own. Poor service quality cannot be blamed or passed off on failings of another support team. The service is suffering no matter what the issue is. The Service Owner works across teams to rectify service issues as needed.

Flexible And Responsive

They recognize that the most important operational strategy is to create a support environment that can be quickly established to meet business needs on a timely basis. They are willing and flexible in support of rapidly changing technologies. They avoid practices that block the ability for the company to deploy new strategies quickly enough to overcome changing market conditions and new competitive threats.

Service Oriented

They recognize that IT operations support is a service versus an administrative body. They aggressively employ processes and communication vehicles to clearly build their service expectations with the rest of the corporation to ensure they meet business needs. They run their operations support just like any service business. They recognize that management needs to understand on-going support costs as part of the overall financial picture of any solution and that they are active participants in finding ways to deliver services at the lowest cost possible.

Priority Driven

They recognize which systems are mission critical to the business and target appropriate levels of support to meet their needs. Their top service priorities typically are 1) Protect company data/application assets with successful backup/ restore and recovery functions, 2) Maximize availability by responding as quickly as possible to isolate and correct faults and problems, 3) Provide an effective problem management and tracking function to ensure business needs are being met, 4) Provide an effective set of firewall processes that guard against unwarranted/ untested changes to technology components. Other support functions are also important but usually follow these in terms of priority.

Proactive Support Providers

They understand that support is not a hand-off or last step in the application development process. This means operational involvement right from the design stage to identify operational requirements and support needs. They recognize that they also "share" in the success of the solution. They provide "operability" value by helping development teams recognize and resolve risks such as how applications will operate once business volumes are in place, whether they scale and perform quickly enough to meet business objectives, escalating alarms/alerts with support services and identifying manual support tasks.

Integrated Solutions Builder

They recognize that support services are not built by technology alone. A total support infrastructure includes all 3 legs of the support solution: Organization, Process and Technology. These need to be considered equally in the service development process to avoid failures, delays in deployment and expending effort on non-critical issues.

Cross-Silo Management Style

They recognize that an operations management structure cannot succeed solely with vertical technical silos. Cross-technology roles that support end-user service needs, identify new application support requirements, assess full impacts of component changes and resolve cross-technical platform problems are needed.

Rewarded For Automation Efforts

The newer systems management technologies in place today provide many opportunities to build automated approaches for support tasks that are manual. Operations support staff are encouraged and rewarded for identifying and building solutions that cut manual effort or reduce overall support costs. This is sometimes implemented via some form of variable pay or bonus initiative.

Measurement Results Used To Improve Not Punish

Concerted effort is taken to measure the services delivered in terms of their quality. Consider an apples to apples comparison of incidents, problems, service breaches and changes across each and every service. High incident counts with low problem record counts might indicate a Service Owner is not being proactive on addressing issues. High incident counts with high change counts might indicate issues with testing changes before they go live. Use the numbers to drive improvements, not to punish staff.

Constant Search For Best Practices and Methods

Ongoing effort is continually taking place to look for the latest technologies and best practices to improve services. This is a proactive activity and it doesn't start at the directive of a CIO or with a business issue. There is a constant search to find ways to deliver services with higher quality and less cost.

The above habits allow you to manage events rather than events managing you. Consider implementing them within your management style. If not, isn't that another fire that needs to be put out?

Descriptions Of Key Service Roles For IT

Service Owner

The service owner is responsible to the customer for the initiation, transition and ongoing maintenance and support of a particular service and accountable to the IT director or service management director for the delivery of the service. The service owner's accountability for a specific service within an organization is independent of where the underpinning technology components, processes or professional capabilities reside.

Service ownership is as critical to service management as establishing ownership for processes which cross multiple vertical silos or departments. It is possible that a single person may fulfill the service owner role for more than one service.

Key activities for the Service Owner Role include:

- Working with business relationship management to understand and translate customer requirements into activities, measures or service components that will ensure that the service provider can meet those requirements

- Ensuring consistent and appropriate communication with customer(s) for service-related enquiries and issues

- Assisting in defining service models and in assessing the impact of new services or changes to existing services through the service portfolio management process

- Identifying opportunities for service improvements, discussing these with the customer and raising RFCs as appropriate

- Liaising with the appropriate process owners throughout the service lifecycle

- Soliciting required data, statistics and reports for analysis and to facilitate effective service monitoring and performance

- Providing input in service attributes such as performance, availability etc.

- Representing the service across the organization

- Understanding the service (components etc.)

- Serving as the point of escalation (notification) for major incidents relating to the service

- Representing the service in change advisory board (CAB) meetings

- Participating in internal service review meetings (within IT)

- Participating in external service review meetings (with the business)

- Ensuring that the service entry in the service catalogue is accurate and is maintained

- Participating in negotiating service level agreements (SLAs) and operational level agreements (OLAs) relating to the service

- Identifying improvement opportunities for inclusion in the continual service improvement (CSI) register

- Working with the CSI manager to review and prioritize improvements in the CSI register

- Making improvements to the service

- Serving as the point of escalation (notification) for major incidents relating to the service

- Representing the service in change advisory board (CAB) meetings

- Participating in internal service review meetings (within IT)

- Participating in external service review meetings (with the business)

- Ensuring that the service entry in the service catalogue is accurate and is maintained

- Participating in negotiating service level agreements (SLAs) and operational level agreements (OLAs) relating to the service

- Identifying improvement opportunities for inclusion in the continual service improvement (CSI) register

- Working with the CSI manager to review and prioritize improvements in the CSI register

- Making improvements to the service.

The table below summarize typical activities that Service Owners may get involved in from an ITSM perspective:

ITSM Area	Role Involvement
Incidents	Is involved in (or perhaps chairs) the crisis management team for high-priority incidents impacting the service owned
Problems	Plays a major role in establishing the root cause and proposed permanent fix for the service being evaluated
Releases and Deployments	Is a key stakeholder in determining whether a new release affecting a service in production is ready for promotion
Changes	Participates in CAB decisions, authorizing changes to the services they own
Configurations	Ensures that all groups which maintain the data and relationships for the service architecture they are responsible for have done so with the level of integrity required
Service Levels	Acts as the single point of contact for a specific service and ensures that the service portfolio and service catalogue are accurate in relationship to their service
Availability and Capacity	Reviews technical monitoring data from a domain perspective to ensure that the needs of the overall service are being met
IT Service Continuity	Understands and is responsible for ensuring that all elements required to restore their service are known and in place in the event of a crisis
Security	Ensures that the service conforms to information security management policies
Financials	Balances cost with value and assists in defining and tracking the cost models in relationship to how their service is cost recovered.

Service Manager

This role oversees the entire IT Service Management (ITSM) operation to ensure that quality service management solutions are developed and deployed to meet agreed business objectives. Key activities include:

- Ownership of IT support for the service end to end

- First point of escalation from Service Desk for incidents and requests

- Monitors work queues for escalated incidents

- Responsible to resolve incidents within set SLA/OLA targets even if they have to sub-task to another service support team

- Responsible for sub-tasking and coordinating activities to fulfill requests that are escalated

- Champions and promotes service improvements on an ongoing basis to continually improve quality and customer satisfaction with their IT service

- Reviews service metrics (KPIs – Key Performance Indicators) that identify the success of the services being utilized to recommend and coordinate implementation of changes to ITSM services to improve metrics

- Ensures continuous alignment of the service with the customers' needs, i.e. changing work patterns, workloads, revised aims and objectives

- Represents the service in CCB (Change Control Board) meetings for Change Management

- Receives change request drafts from support staff, reviews and submits to change management for medium to high risk changes

- Manage support staff and budget for the service.

- Enable and champion an IT service culture.

- Develop, implement and maintain ITIL-based management processes and controls to ensure service quality is maintained to meet business objectives.

- Champion and promote service improvements on an ongoing basis to continually improve quality and customer satisfaction with IT services.

- Maintain day to day responsibility for the ownership and resolution (including any referral or escalation as may be necessary) of Service Management issues which arise in connection with ITSM Services.

- Review service metrics (KPIs – Key Performance Indicators) that identify the success of the services being utilized to recommend and coordinate implementation of changes to ITSM services to improve metrics.

- Work to ensure continuous alignment of the services with the customers' needs, i.e. changing work patterns, workloads, revised aims and objectives.

- Co-ordinate inter-process changes with ITSM process owners.

- Ensure alignment of ITSM solutions to the corporate and IT strategy.

Process Owner

Owns one or more processes and is responsible for process quality and coordinating process with other processes in the organization. Key activities include:

- Responsible for ensuring overall process objectives are met.

- Provides direction to staff using the process.

- Monitors process maturity and progress.

- Measures processes for quality, effectiveness and deficiencies that need to be addressed.

- Coordinates design decisions and activities with other Process Owners.

- Assists in development of project work plans, schedules and staffing requirements from a process perspective.

- Communicates as required to executive management and Program Office.

- Ensures that process implementation and design requirements are adequately identified and that process solution issues are being addressed.

- Identifies process and solution requirements to Technical Architecture Teams.

- Ensures people are using the process effectively.

- Coaches and teaches others about process concepts and solutions.

- Communicates the process throughout all support teams.

- Provides overall leadership and management from a process perspective.

Business Relationship Manager

This role provides a single point of contact to one or more business units for all IT services. It reviews quality of services delivered to those units, addresses customer issues and communicates changes in customer needs back to IT. This role is very critical. Here are some examples of what can happen when this role is not in place:

- Services will still be delivered and can meet delivery targets but IT cannot quantify the value of the services

- There is no guarantee that appropriate business unit needs are being fully met

- There is no guarantee that services and service investments are being prioritized correctly

- There is no guarantee that the customer IT meets with are truly representing the business needs of the customer

Service provision without business relationship management is possible, but can make delivery of services costly, erratic and foster mistrust between IT and their customers.

Key activities for this role include:

- Establishes and maintain a business relationship between IT and one or more business units based on understanding the customer and their business needs.

- Identifies business unit needs and ensure that IT is able to meet these needs as business needs change over time and between circumstances.

- Ensures high levels of business unit satisfaction, indicating that IT is meeting customer requirements.

- To establish and maintain a constructive relationship between IT and the business unit based on understanding the customer and their business drivers.

- Acts as a single point of contact for one or more business units for all IT services

- Identifies business unit IT needs and helps the business identify the services and options that best meet those needs

- Communicates business unit service issues and requests for new services back to IT

- Communicates service status on service issues being addressed by IT back to the business unit

- Documents requests on behalf of business units for complex and unique IT services and requirements not in the IT Service Catalog

- Assists in service negotiation efforts between IT and the business unit representing the "Voice Of The Customer"

- Assists in coordinating service improvement projects, transition projects or other related efforts specific to one or more business units

- Reviews quality of services rendered to the business units being represented on a regular periodic basis

- Escalates business unit service issues to the IT Program Office

- Documents requests on behalf of business units for complex and unique IT services and requirements not in the IT Service Catalog

- Communicates service status on service issues being addressed by IT back to the business unit

- Assists in service negotiation efforts between IT and the business unit

- Assists in coordinating service improvement projects, transition projects or other related efforts specific to one or more business units

- Reviews quality of services rendered to the business units being represented

Some considerations and challenges to watch out for when implementing this role are:

- Staying actively involved - ensuring that role is not perceived as "window dressing" just to make IT customer friendly

- Handling customer satisfaction when the services delivered by IT are less than optimum

- Focusing on the business unit relationship without becoming mired with detailed requests or acting as a "go-fer"

- Balancing the "Voice of the Customer" even though this role may work directly for IT

- Handling the situation where a business unit customer (the one buying the services) does not set expectations with business unit users (who may have expected higher levels of service than what was purchased)

Technology Owner

Owns one or more technologies and provides a single point of contact for them. They are responsible for proper operation of those technologies, how they are used to underpin services and also provide information on future direction of the technologies owned. Key activities for this role include:

- Provides single point of ownership for effective provision of technical support services to customers and stakeholders.

- Oversees technical support activities and services provided to support teams.

- Oversees design tasks related to development of new systems and services.

- Manages and develop all technical support staff for the service.

- Oversees recruitment of technical support staff.

- Assigns service technical support staff to projects and initiatives as needed.

- Resolves technical support issues and design decisions that have been escalated.

- Analyzes technical issues.

- Resolves infrastructure incidents and problems when they occur.

- Maintains ownership of problem diagnosis, resolution and escalation for all received problems and issues.

- Participates in service design and transition technical activities for new or changed services when needed.

- Applies third party maintenance to infrastructure components.

- Develops and maintains infrastructure documentation and procedures.

- Provides technical support and assistance with feasibility studies.

- Responsible for all uses of the technologies owned.

- Responsible for ensuring that appropriate skills are in place to install, build, maintain and support the technologies owned.

- Monitors technologies for proper operation.

- Coordinates design decisions and activities with other Technology Owners.

- Oversees technical support activities for technologies owned.

- Assists in development of project work plans, schedules and staffing requirements from a technology perspective.

- Ensures that implementation and design requirements are adequately identified and that technology solution issues are being addressed.

- Identifies technology requirements to Technical Architecture Teams.

- Ensures technologies are being maintained to vendor specifications.

- Ensures technologies are adequately supported by vendors.

- Coaches and teaches others about concepts and solutions concerning the technology.

- Identifies and manages to a future roadmap for technologies owned such as for new technology releases or advancements in functionality.

- Provides overall leadership and management from a technology perspective.

ITSM Program Manager

Unlike the other roles which are used to operate IT Service Management on an ongoing basis, this role exists to manage and lead one time major IT Service Management initiatives. The role may be necessary for situations such as a large IT Service Management transformation effort or other significant effort that will involve significant organizational change. The role itself is a combination of project/program management and organizational change leadership. It owns the transition initiative end to end. Key activities can include:

- IT Service Management vision and planning.

- Project and program management across the many transition efforts and implementation teams.

- Program leadership across technology, process and organizational change projects and activities.

- Budget management and reporting for the ITSM initiative.

- Acts as an escalation point for key ITSM decisions and issues.

- Recruitment of ITSM program team members and support staff.

- Problem resolution for key program issues that cannot be resolved at lower levels.

- Approval or recommendation of key ITSM solutions, operating architecture and policies.

- Acts in the role of primary approver for key ITSM technology purchases and 3rd party support contracts.

- Reports overall program status and progress on key issues to IT and business executives.

- Retains primary accountability for the overall ITSM Transition program and its success.

In short, this role serves as the right hand person to IT and business leadership for getting the ITSM Transition program completed. Once in place, this role can morph into an ongoing Service Management director or oversight role.

Recommended skillsets for this role can include:

- Ability to create a culture of innovation

- A strong sense of customer service measurement, process design, and customer service behaviors

- Leadership ability to motivate staff toward high visions

- Understanding of the application of organizational change and group techniques to move people towards the ITSM vision

- Leadership experience with tool and process development projects

- Deep project and program management skills

- Adept in running and growing self-managed teams

- Adept at charter development and process analysis

- Strong meeting facilitation skills

- Competent in measurement development, scorecards, dashboards and reporting to executives

- Able to define performance and service gaps with existing operations as well as for proposed solutions

- Strong communication skills

- Adept in process analytical skills, process maps, flow charts, and other analytical tools

- Missioning and visioning development skills

- Strong ability to mentor others

ITSM Organization Models

Overview

Organization: a group of people intentionally organized to accomplish an overall, common goal or set of goals. Business organizations can range in size from two people to tens of thousands.

Organization Design: a formal, guided process for integrating the people, information and technology of an organization. It is used to match the form of the organization as closely as possible to the results the organization seeks to achieve. Through the design process, organizations act to improve the probability that the collective efforts of members will be successful.

IT organizations come in many styles. There is no right organization chart or magic organization structure that can satisfy all needs. The structure is generally chosen by the industry and business mission of the company. Examples of different kinds of structures as they apply to IT organizations might include those as shown below:

Functionally Based

The IT organization is organized by operating functions or related sets of activities. Examples might be the Service Desk, Operations Management, Technical Management, Application Management functions.

Product Based

The IT organization may be aligned with specific internal products like Windows, LINUX, Oracle, and SAP platforms. Alignment may also be to external products that the business directly sells to their customers.

Service Based

The IT organization may be aligned with specific services that offered internally and externally to business customers. Examples might include support services, maintenance services or hosting services.

Customer Based

The IT organization may be aligned to very large or important customers that the business serves. One example might be an IT support organization that is local to a particular business unit. Another example might be an IT support organization that only services one or two business customers.

Geographic Based

The IT organization may be aligned by different geographies. One example might be an IT support organization that has separate structures broken out by Americas, Europe, Middle East and Asia.

Business Process Based

The IT organization may be broken out by business processes. One example might be an IT support organization that is broken out by Accounting, Payroll, Legal and Human Resources.

The above styles don't necessarily have to be one or the other. Many times IT organizations will choose to combine and mix each of these as needed to meet business requirements.

IT Organizational Models

Below are examples of common forms of IT organizations. There is no one right answer, however pros and cons of each one described are presented.

Decentralized Model

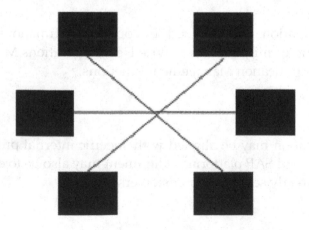

With this model, there are separate IT organizations each led independently by a CIO. There is no overall control. Actions are coordinated by contracts and agreements rather than through a formal hierarchy.

Pros:

- Fastest adoption approach - each IT unit can adapt quickly in a way that benefits them best

Cons:

- No overall control across IT organizations
- May have variation in process across IT units
- May have duplication in assets and resources across IT units
- Collaboration across IT units will be ad-hoc
- Little incentive for one IT unit to assist another
- Accountability for IT service coordination is at the executive level

Best Use:

- IT units that support independent businesses with separate P&L (profit and loss) accountabilities
- IT units that support independent industries where there is no need for cross industry support

Centralized Model

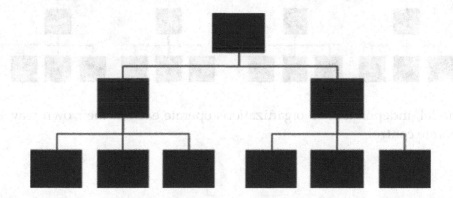

With this model, all IT functions are provided by a single organization which all IT executives, middle management and staff report into.

Pros:

- Provides clear top down direction setting
- Speeds decision-making processes
- Service accountabilities are very clear

- Provides efficiency for internal activities

Cons:

- Impedes service delivery efficiency since key decisions are dependent on the top

- May stifle innovation and entrepreneurship

- May create barriers for trying to work closely with customers that have special needs

Best Use:

- IT organizations under severe cost pressures or constraints

- IT organizations serving business units that also are highly centralized (e.g. government)

- IT organizations that support high risk business functions or customers

Corporate Directional Model

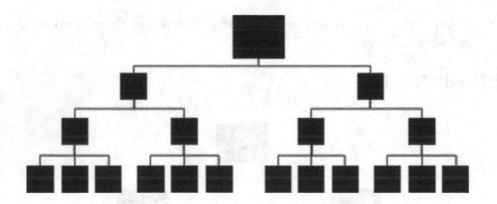

With this model, independent IT organizations operate each in their own way, but under overall corporate control.

Pros:

- Provides overall control through policies and standards and possibly measurements

- Allows each IT unit to implement in their own way

- Relatively fast to adopt

Cons:

- At risk for variation across IT units unless tight measurements and controls are in place

- Leaves accountability for service coordination at the corporate level

- IT units may have to fight for competing needs

Best Use:

- Very large, multi-national, IT organizations where there is single leadership at the top

- Delivering services across differing regions or countries where local variation is needed

Centralized With Local Variation Model

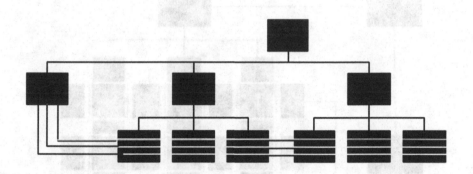

In this model, separate IT functional units exist, but an additional unit has been added with direct report into the CIO accountable for aligning IT activities across those functional units. Formal systems (Hint: IT Service Management) used to coordinate activities must be in place.

This model is especially good for ITSM transitions where the added unit can coordinate ITSM activities without disrupting the IT line organization.

Pros:

- New unit can have accountabilities for unique needs such as ITSM processes and services or other initiatives

- Allows other IT units to focus on their core service delivery capabilities while transition to new processes and services is taking place

- Lowers risk of variance in processes used across all IT units

Cons:

- Process and service issues when they arise will have to be addressed at CIO level or through a governance council

- Must work through cross matrix issues as the additional unit works with other IT units to resolve conflicts in priorities

Best Use:

- Transition to major IT solutions and methods such as ITSM

- Implementation of new initiatives or controls that need to span across all IT units

- Implementation of specialization of products and services for specific customers or business markets

- Provision of specialized skill sets that would be needed across all IT units

Collaborative Model

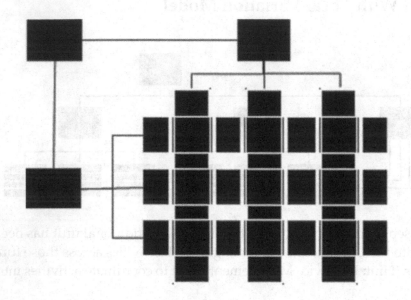

With this model, a virtual organization is created within the existing IT units creating a multidivisional matrix structure. This provides for more integration between corporate and divisional IT units and between divisional IT units. One example is to have the vertical units own technologies while the horizontal units own customers and services.

Pros:

- Easier for top executives from the IT divisions and from corporate headquarters to coordinate organizational activities

- Adaptable for developing products while still meeting customer specific needs

- Minimizes organizational change

- Can provide accountability for processes and services if done correctly

Cons:

- Potential for conflicts between responsibilities for service management versus individual IT unit needs (or between technology and customer groups)

- Will still require overall leadership and coordination

- Requires high levels of teamwork

- May slow down delivery of products and solutions due to needs for coordination and agreement

Best Use:

- IT service delivery where a "high touch" level of service is desired with customers

- Ensuring products sold to customers are meeting actual customer needs and desires (versus taking customer surveys or other activities where the product organization has to interpret what they think customers really want)

- Desire to truly instill service quality at an equal level with IT technologies

Chapter
5

Building The IT Organization

"Make changes WITH people – not TO people"

This section presents one approach that can be used for building and designing an IT organization. Note that you can't just focus on structure and job positions. There needs to be a strong understanding of the services the IT organization is delivering (or about to deliver) and the support and operating processes that underpin those services. Without this, you are just rearranging the deck chairs on a ship which may yield little or no better results than what was being achieved previously.

Triggers for Changing the IT Organization

Both external and internal events can trigger the need for an organization redesign. Internal factors can include:

- A new company or division is being added to the business that requires additional IT services and support

- The business is planning to grow such that the current IT organization may not be able to scale with that growth

- The business or IT strategy has changed

- The organization around the business has changed as a result of a merger, acquisition or divestiture

- The current IT organization isn't delivering the performance expected

- There has been a change in senior leadership such that new leadership desires changes in how IT is organized and operated

- There is a desire to outsource or insource IT functions of the company

Examples of external factors that could trigger changes in the IT organization can include:

- Changing competitive landscape for the business

- Major new customers or changes in how customers are to receive their IT and business services

- Company push towards new markets and geographies

- Regulatory changes that have wide impact on how the business or IT should operate

- Acquisitions, mergers or business consolidation

- Severe business issues such as major outages or a downturn in business conditions

- Company desire to:

  - Add or remove a level of management

  - Increase or decrease span of control

  - Centralize, decentralize, consolidate or separate business functions

  - Align structures with processes or customers

Principles for Organizational Design

Some design principles to consider as you go about this task are as follows:

Balance supply and demand of resources for every organizational unit and individual.

The supply of resources is determined by the level of control over resources as well as the level of support that an organization unit or individual has. The demand of resources is based on the accountability assigned to organizational units and the degree of influence an organizational unit/individual has to exercise to meet their goals.

Balance the tension between "differentiation" and "integration."

The need for differentiation to allow for economies of scale and specialization while providing adequate integration to allow for coordinated work and realization of interdependent goals

Find the structure and roles that will process information most efficiently.

The design should build the capacity of the organization to process information in keeping with the level of uncertainty faced by the organization. Higher levels of uncertainty require higher information processing capacity.

Organizational Balance

How to balance the formal organization structure and performance measures with the informal workgroup dynamics and how people will work together.

Decision Hierarchy

Determining what will be the most efficient and effective decision levels and how decisions get escalated. What kinds of decisions will be made by senior executives versus middle management versus line staff.

Structure

Determine the organizational model or set of models that will be used for reporting and control. (Example models were presented in the earlier section)

Roles and Accountabilities

Determine which roles will be needed to carry out and support business activities. What activities will each role execute? Which roles will be accountable and responsible for those activities?

Positions and Skill Sets

How roles will be grouped into job positions. What skill sets will be needed in each job position.

Performance Measures

What measures will be assigned to each job position that will best drive desired behaviors.

External Factors

What external factors exist that may put constraints on the organization. Examples of these might include considerations for labor unions, regulatory or legal factors.

Structural Integration

Structure greatly affects how employees will act, interact and behave as well as the type and degree of supervision that they need. Work needs to be logically focused, designed and organized. The target structure is key towards determining the efficiency, quality, cost and amount of work that can be performed in a given amount of time.

Other factors for consideration can also include:

- Impact of the existing culture on the target state design

- Degree of self-directedness that exists

- Target deadlines or timelines for when the new organization needs to be put into place

- Whether current staff can be trained for new job positions versus hiring to fill skills gaps

- Existing labor agreements that are in place

- Whether the business is growing or contracting

- Whether performance measures exist and if they are compatible with the desired design

- Whether services will be outsourced, insourced or put into the Cloud

Designing the New IT Organization

The following is a suggested approach that can be used to design an effective IT organization:

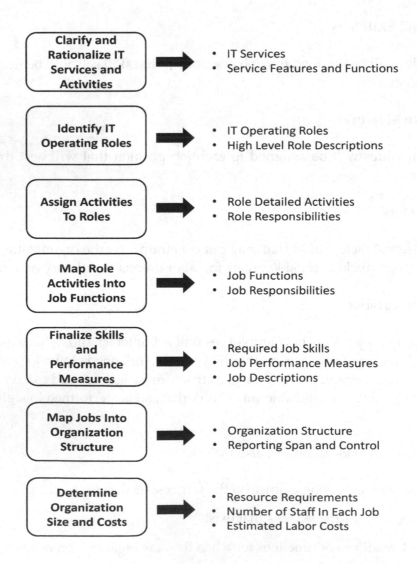

Before starting, there is a key question. Which option for organization structure do you intend to take?

Option A

Make no organization changes. Use the current IT organization structure as is.

Option B

Use the current IT organization structure but tweak it a bit with some changes in reporting structure and alignment.

Option C

Redesign the IT organization structure into something totally different.

Option D

Outsource some or all of the organization jobs and job activities to a third party.

Some design activities may differ depending on which option you are going after. A more detailed description of each step follows.

Step 1: Clarify and Rationalize IT Services and Activities

This is a key step in getting started. It's hard to design an effective IT organization if there is lack of clarity in what IT services are being delivered. The Servicing ITSM book provides detailed guidance and a starting set of services for doing this.

The output of this step is to ensure that a solid set of high level IT services, service descriptions and features are in place before starting. If these are already in place, then this is an easy step to simply confirm them. Without this being in place, the design effort may become a pointless exercise in rearranging existing activities to accomplish some unknown end.

If you use Option D, then identify which services are to be outsourced to third parties.

Step 2: Identify IT Operating Roles

In this step, determine what logical roles will be used to operate the IT organization. Roles are not to be confused with job functions. Roles are logical groupings of related activities. These should be designed independently of any existing job functions or organization structure.

If not sure where to start, the ITSM Operating Roles chapter in this book provides a nice starting set of operating roles that you can use or modify for your own needs.

Step 3: Assign Activities to Roles

In this step, determine the activities that each role will contain. A starting set of these for each ITSM operating role is presented in the ITSM Operating Roles chapter of this book.

Step 4: Map Role Activities into Job Functions

If you chose Option A, then simply map each activity identified in Step 3 into a job function that already exists. When doing this, ensure that all existing as-is activities have been included along with each job function.

For Option C, you will need to first identify a set of high level job functions. Job functions will become actual physical jobs made up from the role activities designed earlier. Options for building these can be:

Take each role and simply make it a job function. (1 to 1)

> In this case, the role description and all the role activities will be the job function. For example, the role of Service Desk Agent is simply now the Service Desk Agent job.

Combine multiple roles into a job function. (Many to 1)

> In this case, take multiple roles and combine all their activities into a single job function. For example, all the activities in both roles of Service Desk Agent and Request Analyst will be combined together for the Service Desk Agent job.

Split roles into multiple job functions (1 to Many)

> In this case, take each role you have defined and place it in two or more job functions. This can be done by splitting the activities into separate jobs. For example, the role of Service Desk Agent may be split into two jobs. Some activities will be part of a Service Desk Tier 1 job, the other activities might be part of a Service Desk Tier 2 job.

Split activities from multiple roles into multiple job functions (Many to Many)

> In this case, take all the activities in two or more roles and map them into two or more job functions. For example, the roles of Service Desk Agent and Request Analyst may have subsets of their activities split into multiple jobs such as the Service Desk Analyst job, Service Desk Call Agent Job and the Request Fulfillment Agent job.

Outsource Roles and Activities

> In this case, combine all the activities from each role to be outsourced into requirements for selecting an outsourcer.

For Option B, you will use a hybrid where some activities go into existing IT jobs while others may be combined into new IT jobs.

Step 5: Finalize Skills and Performance Measures

For this step, take each job function and identify the needed skills to operate it. Jobs need both a skill and a skill level. For example, does the Service Desk Agent job need to have an expert level of technical skills or maybe just enough to solve common technical issues? Does this job require a basic set of customer relationship skills or something more advanced?

The ITSM Operating Roles chapter of this book provides a starter set of operating skills and suggested skill levels that you can use to get started if you need additional help.

Develop a set of quantifiable metrics that can be used to determine and assess employee productivity. For performance measures, develop functional competency models that create a link between employee skills and organizational goals to measure performance. Determine what measures will be needed to recruit staff as well as those needed as a way to guide employee development or link to a pay and bonus structure. For this activity, human resources will need to be more involved.

Step 6: Map Jobs Into Organization Structure

The main goal of this step is to map the jobs into the target structure. If Option A was selected, then this might result in tweaks and changes to current job descriptions and some responsibilities.

For options B and C, develop organization design scenarios utilizing previously defined design objectives and criteria and the operating model(s) that will be selected. Ensure that all Jobs are accounted for in the design.

For Option D, identify which jobs may stay in house versus those that are outsourced. Note that with this option, it is possible that some additional activities may arise for needing to interface and deal with the outsourcer once selected.

Step 7: Determine Organizational Size and Costs

In this step, determine final size of organization by job function and geography. As a final step, estimate the labor costs for each job function. Develop a cost model from this that can be used as a business case and to communicate the organization design with executives.

Building RACI Models

For an ITSM delivery organization to be successful, it is essential to define the roles and responsibilities within that organization to execute on the various activities. A popular approach for defining organizational roles and responsibilities is the RACI model.

Assignment of responsibilities to various roles should not be left to chance or done at the last minute. If that happens, there is potential for conflicts and finger pointing between staff members. Decisions are slowed down. A means needs to be provided to identify who is responsible for what ahead of time so those issues do not get played out in front of a customer, during a service outage, or in the course of delivering customer services.

The letters RACI stand for **R**esponsible, **A**ccountable, **C**onsulted or **I**nformed. The model identifies service delivery roles, lists service delivery activities and then maps them against each other using one or more of the letters to show the level of responsibility for that activity. The only key rule is that one, and only one, role can be accountable for an activity. This is to ensure that there is no confusion as to which role is in charge of any activity.

Scope of a RACI model can be anything from the entire ITSM delivery organization and its users down to a single process or activity.

A spreadsheet makes a good tool for developing a RACI model. The steps for building it are as follows:

Step	Action
01	Identify the activities and list in each ROW
02	Identify the roles and list in each COLUMN
03	Assign the RACI codes in each intersecting CELL
04	Ensure every ROW has only one A in it
05	Ensure every ROW has at least one R in it
06	Agree the chart with key stakeholders
07	Publish the chart to stakeholders
08	Make sure allocations are followed

More detail about the various allocations and their rules is as follows:

RACI Role: Responsible

- The role that performs the action/task

- The scope and degree of responsibility is defined by the Accountable role

- Responsibility can be shared across multiple roles

- Responsibility can be delegated to other roles

- A role can be both Accountable and Responsible if necessary (as long as no other role is also Accountable)

RACI Role: Accountable

- The role that is held accountable for completing the action/task

- The role that has ultimate accountability and authority for an action/task

- There is only one accountable (A) to each task/activity

- Accountability is assigned at the lowest level and implied at higher levels

- Accountability cannot be delegated

RACI Role: Consulted

- The role that is consulted before performing the action/task

- The role consulted prior to a final decision or action being taken

- May provide expertise, opinions, facts or other information to the other roles

- Two-way communications exists between this role and other roles

RACI Role: Informed

- The role that is informed after performing the action/task

- The roles that need to be informed after a decision or action is taken

- May also be roles that are informed an action or task is about to be taken

- Provides no decisions or actions other than getting informed

- One-way communications exists between this role and other roles

An example of a RACI model is presented below:

Activity		Service Level Manager	Service Level Analyst	Service Owners	Process Owners	Business Liaison
1.2.1	Identify services	A/R	C	C		C
1.2.2	Manage expectations	A/R	C	I	I	R
1.2.3	Plan service level agreement structure	A/R	C	C		
1.2.4	Establish requirements & draft SLAs	A/R	C			C
1.2.5	Negotiate service level agreements	A/R	C	C	C	C
1.2.6	Establish monitoring capabilities	C	A/R			
1.2.7	Review UCs and OLAs	A/R	C	C	C	C
1.2.8	Define reporting & review procedures	A/R	R	C	C	C
1.2.9	Publicize SLAs	A/R	C	I	I	I
1.3.1	Monitor service levels	I	A/R	I	I	
1.3.2	Report service	I	A/R	I	I	
1.3.3	Conduct service review meetings	A/R	C	C	C	C
1.3.4	Manage service improvement program	A/R	C	C	C	I

Organizational Transition

The following is a suggested approach that can be used to transition to the target IT organization:

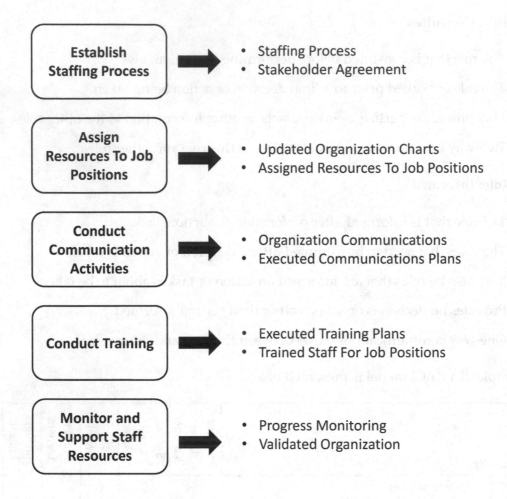

Step 1: Establish Staffing Process

In this step, establish the process for staffing the organization. Work closely with and adhere to HR/Legal guidelines that may be needed throughout the staffing process. Develop HR resource plans that are consistent with any overall HR talent management programs. Ensure all stakeholders are prepared to execute against plans.

Step 2: Assign Resources To Job Positions

Activities in this step will fill and/or transfer employees into the job positions. A suggested set of activities for doing this include:

- Identifying which employees will fill which job positions

- Identify timeframes for when those positions will be open and filled

- Developing information and activity packets that can be used by leaders to better facilitate the transition to the new organization

- Completing new organization charts with names disseminated organization-wide

- Formally notify selected employees of their new positions in the new organization

- Formally notify non-selected employees at the conclusion of people selection

- Ensure that the staffing process is conducted according to plan

Step 3: Conduct Communications Activities

In this step, the communications plan is executed to transfer employees into their job roles (or change activities performed within existing job roles if those roles will still exist). Activities in this step can include the full set of communications events as will be shown in later chapters of this book. While this section is focusing mostly on Organization structure implementation, some other considerations can also include:

- Timely communication with newly selected employees by management after each level of selection

- Holding formal announcements for newly selected employees to the organization

- Welcoming newly selected team members

- Communicating expectations for the new organization

- Tracking and monitoring progress on resource assignments

Step 4: Conduct Training

In this step, plans are executed to train employees into their new job roles (or for handling new activities performed within existing job roles). Much more on training and training strategies is presented later in this book. Other considerations relevant to organization structure include:

- Validating that employee skills and knowledge are at acceptable levels to more effectively execute job responsibilities

- Monitoring training attendance and progress to ensure training activities are being carried out and completed on time

Step 5: Monitor and Support Staff Resources

In this step, monitor staff participation in their new job roles (or with changed activities if roles did not change). Conduct ongoing activities to support staff in their new roles and address any issues. Finally, assess how well the new organization is meeting desired benefits.

Dealing With Resistance

Resistance is a natural reaction to change – don't be surprised by it"

Basic Causes for Resistance

Recognize that it is normal for people to hate change. We are all programmed by nature to act in this way. This goes back to our early cave-person days where we were programmed to act suspiciously of any change in weather, neighboring tribe, or hunting situation as a survival mechanism. In addition, those of us who are older in IT have spent a lifetime doing things a certain way with a reasonable amount of success in the past. Why bother to change now?

From this, you should recognize that people go through different stages in accepting change. They may be initially suspicious or hesitant to do things differently, but with the right exposure can adapt over time. The trick of your ITSM Communication Strategy is to provide the right exposure to overcome these kinds of barriers and gain acceptance.

Think about these scenarios:

Scenario #1: The ITSM Transformation team goes offsite and develops ITSM processes and procedures. They then document these, return to the work site and get management to immediately make sure everyone follows what was documented.

What do you think the chance for success will be in the above scenario? Let's look at another:

Scenario #2: The ITSM Transformation Team drafts a Vision for how IT should operate. They then review this with key business and IT stakeholders. While solutions are being developed, a series of brown bag sessions and seminars are held over time. The first merely states why this is important. The next one expands upon this and present a few additional concepts. The following session involves a little game and role playing. Following this, people are trained in key concepts. Then an internal certification program in ITSM solutions is conducted. Over time, ITSM solutions are brought in piece meal and people are given a chance to provide feedback on their use and success.

In the first scenario, there is no preparation for change. Everyone is told to just "do it". There is a high risk that there will be much resentment among staff doing something "someone else's way".

In the second scenario there is at least a shot for success. People were introduced to new concepts over time and in small chunks. They had a chance to try out solutions ahead of time. They could even provide feedback and recommendations to what was put into place. They feel part ownership of the solutions and will therefore tend to not resist them.

This leads to several important tips when working on ITSM transformation efforts:

- People respond better if they had a hand in building/recommending the solution

- People respond better if introduced to new concepts in small chunks over time

- People respond better if allowed to experience new things in the form of role playing or game playing

Another consideration is to realize that people accept and process change in many different ways. This is somewhat like the stages people go through when told extremely bad news (like you're about to die):

1. Denial ("Won't happen to me" or "Just another IT fad – it will soon pass")

2. Fear ("What does this mean to my job? Will I no longer be top dog around here?")

3. Pity ("Why are they doing this to me? Don't they know how hard my job already is?")

4. Bargaining ("Okay, but I'll only submit changes for applications – not infrastructure…")

5. Acceptance ("Great! I'm in!")

Another way to recognize and categorize resistance might be summarized by the following chart:

Behavior	Issue	What They Say	What They Do
Complacent	People do not understand that there is a need for ITSM	We are the best Nothing can stop us Don't fix what isn't broken We know what we are good at We've always been successful	Pat each other on the back Get less cost conscious Take their eye off the ball Promote past history of IT
Denial	People hope ITSM will go away or do not believe it is theirs to do	This is the fault of someone else We can't do that because… This happened before and we got by okay… Good point but you don't understand…. Our business is different We're too busy with other priorities	Finger-point and blame Become overly optimistic Reject key facts, metrics Pass the buck Look to past achievement
Confused	People understand ITSM but not how to implement it	How did we get in this mess? Where is this all going? Who, what, where, why…? Too many initiatives around here Hire a consultant?	Look for direction Don't fully complete everything Adopt the latest fads Endless speculation and rumors Generate lots of overlapping or conflicting initiatives Constantly jumps into whatever sounds like the Holy Grail…
Busy	People are too busy with other priorities to focus on ITSM	Don't involve me until absolutely necessary We're too busy to focus right now on ITSM We have other priorities right now	Find excuses to not participate Avoid showing up at meetings Assume this is a fad that will soon pass

Learn to recognize these. People can accept change if given a chance over time. Realize that those who initially resist may become more accepting if given the right levels of exposure over time. They may follow the stages of acceptance as described earlier. Understanding where people are coming from can greatly help you lead and overcome resistance when it rears its head.

5 Stages of Resistance

Unfortunately, nature has programmed us as human beings to detest change. Much of this goes back to our caveman days. We all want to learn how to do something, and then continue to do it the same way for as long as possible unless confronted with a stark reality that it no longer works (in the caveman case, a food source that is no longer working). At that point we go into panic mode until we find another working solution and then repeat the behavior.

IT has seen this happen in an accelerated mode. The number of new IT initiatives, new working approaches, new technologies, different ways of working cycles continuously. Our response each time something new comes along tends to initially meet with resistance. This resistance typically runs in 5 stages progressing through each one until acceptance occurs. These are described below:

Resistance Stage 1: Denial

Everyone starts here. The new change is announced and we all immediately jump to the "hope it never happens" or "probably doesn't impact me" kind of thinking. This can best be dealt with by communicating over and over to create a sense that the change is for real and just a matter of time.

What they usually say:

- "Ignore it – it will pass"

- "It will never happen here"

- "We're different – this doesn't apply to us"

- "It's not important enough to focus on right now"

What you can do:

- Communicate over and over to create a sense that the change actually will happen

- Structure communication events to deliver things in small chunks of information making sure you hit each stakeholder multiple times

- Get executive leadership to speak for just a few minutes about the change at key meetings and events to communicate why the change is important to them and good for the organization

- Ensure your ITSM project is creating some initial wins and delivering per the advertised schedule – delays and overruns will reinforce the feeling that if they continue to ignore the solution it will go away.

Resistance Stage 2: Anger

Once past the Denial stage, people are now convinced that the change will happen. At this point, they become irritated as a result of not fully knowing what the new solution is or how it will be put into place. Lack of answers can contribute to anger. Providing lots of information and addressing points of misunderstanding and confusion will help people get over this stage.

What they usually say:

- "I'm tired of seeing communications about this" (e.g. emails, meetings)

- "Your team doesn't know what they are talking about"

- "This is not the right software product to use"

- "Your process is missing things that are important"

- "This solution is not ready for us to use"

- "This will slow everything down"

What you can do:

- Answer concerns and issues with facts keeping emotion out of the discussions

- Make sure you provide a channel for people to air their grievances – sometimes they just need to hear themselves talk a bit

- Make sure that people are not operating with misunderstandings for how things will really work

- Make sure you truly address concerns and issues when raised – ignoring them only festers the anger and increases the resistance

- Sometimes people just want to express their ideas and even participate in some way – if blocked from doing this, they usually respond with anger

- Let people know that they have a period of time to express their anger, fears and concerns – after that things will move forward

- Allow some time for expressions of anger - after that, let people know that "if you have something to say then say it – else starting in month XXX we're moving forward with this"

- Don't let anger fester or go under the radar – it will pop up again later

- When allowing expressions of anger, make sure the forum is such that it doesn't convey negative perceptions about the solution to the rest of the stakeholders.

Resistance Stage 3: Fear

At this point, you have probably accomplished the first key task in communications. People know the change is coming. They understand what it is that will change. You've communicated the right side of the brain but not the left. This is where fear sets in. There is a conflict with their desires, hopes, what they think is right in the workplace. What will this mean to them personally? Overcoming this stage involves giving people time to air their concerns and personal issues with the change.

What they usually say:

- "Will this expose my deficiencies?"

- Will I have to work much harder?

- "Will I be held to a much higher standard?"

- "Is my job in jeopardy?"

- "Is my promotion in jeopardy?"

- "Can I be successful at this?"

- "Do I have to rethink my career strategy?"

- "Can I be successfully accountable anymore once the new system is in place?"

What you can do:

- Answer concerns and issues with facts keeping emotion out of the discussions

- Make sure a channel is provided for them to discuss their fears and concerns

- Be up front about what is changing and allow them to ask questions

- Watch out for rumors and other falsehoods being sure to stop them as soon as possible

Resistance Stage 4: Bargaining

At this stage, they are close to acceptance, but not sure if they want to fully engage or engage just enough to meet management requirements. Helping them get past this stage involves publicizing ITSM accomplishments and communicating the benefits that can be achieved when everyone is full in on the solution.

What they usually say:

- "I'll support change management for application changes, but not for infrastructure"

- "I expect more money if my role is to change"

- "I can't participate, but will provide someone else who will"

- "We'll do this for servers and applications, but not the network"

- "This will only apply to the infrastructure team, but not the rest of IT"

What you can do:

- Reinforce the ITSM benefits and why they can't be achieved unless everyone is participating

- Look and solve for root causes that contribute to not participating fully (e.g. time issue, resources, skills, and discomfort with the new system?)

- Reward for participation and demote for non-participation (some IT organizations will limit promotions only to those participating)

Resistance Stage 5: Acceptance Reversal

Finally at this stage they have accepted the change. They engage and participate but may slowly slip back to old habits over time. The reversal can be caused by many factors such as a change in leadership, change in ITSM teams, company reorganization, merger or acquisition. It can happen so slowly, no one notices until things start to go wrong. Deal with this stage by continually monitoring ITSM results and ensure that a Continual Service Improvement (CSI) program is put into place.

What they usually say:

- "We used to have good management practices around here about 4 years ago – not sure what has happened to us since then"

- "Can we do it the old way just this once?"

- "Things seemed to go well until the ITSM team moved on"

What you can do:

- Publicly recognize compliance with the new ITSM program on a periodic basis – not just once after implementation

- Publish service improvement results on an ongoing basis and make them public

- Make sure a Continual Service Improvement (CSI) program is in place

- Hold regular meetings post transition to discuss what is working and what is not working

- Conduct periodic assessments and audits to measure ITSM compliance and improvement

Types of People and How to Deal With Them

Tooling Maverick

Characteristics:

- Starts with procuring a tool first
- Feels process should be done by another team disconnected from the tooling team
- May feel that technology can solve almost anything

Outcomes if left alone:

- Implements services, processes, process categories, data, request flows to make the tool work without consulting others
- Lack of preparation has vendor resources burning labor hours usually leaving a "sub-optimal" implementation behind
- Track record shows a high level of failed implementations or rework to reinstall tools with correct processes, services and workflows

Key Strategies:

- Make sure they work under a manager that is already bought into the whole solution not just the technology
- Provide recognition for the tooling contributions this person can make
- Identify case examples where the tool can fail if the wrong processes and services are not input correctly
- Establish a working relationship where requirements are developed first and then put into the tooling solution

The Assessor

Characteristics:

- Hires outside consultant to assess IT processes for maturity levels
- Selects processes to assess – usually incident, problem and change
- Scope focuses almost entirely on processes leaving out tools, organization review of operational artifacts and governance

Outcomes if left alone:

- Creates a report outlining 200+ items that "are not ITSM standard"

- Focuses efforts on how to improve process maturity scores versus solving real IT and business problems

- Raises management interest in maturity score results and sets targets for higher maturity scores

- Leaves others to figure out what the real business issues are and how any of these will be addressed by the process improvements cited

Key Strategies:

- In addition to maturity scores have them identify what business issues and risks exist

- Have them suggest strategies and solutions to address those risks

- Try to broaden their focus to include technology and organizational issues that also need to be addressed

- Have them concentrate on the top 5-6 issues that must be addressed versus all ITSM processes

The Trainer

Characteristics:

- Trains and certifies selected IT staff in ITIL – usually at the Foundation level

- May struggle to relate training concepts to real world application and how ITSM can directly solve IT business issues

- Advice and recommendations usually offered in broad terms that are generally correct but not tactical enough to move forward with

Outcomes if left alone:

- Selected staff get certified and familiar with ITSM concepts

- May schedule training events well ahead of when people actually need to apply what they have learned

- May leave everyone adrift trying to figure out what to do next

Key Strategies:

- Leverage this person's skills to develop and conduct training

- Place in a program support role versus a leadership role unless actual leadership skills are demonstrated

The "Busy" Executive

Characteristics:

- Believes that best practices are a good thing but too busy to participate beyond appointing someone else to implement them

- May see ITSM as a solution for all IT issues

- Places team under a technical support or operations manager (they get all the incidents don't they?)

Outcomes if left alone:

- Leaves ITSM team with little capabilities, support or leverage to work cross organization

- Appears surprised when resources and effort are needed beyond people's day jobs

- De-prioritizes ITSM efforts and starves the team from getting any resources

- Support staff continues to prioritize fire fighting over prevention

- High risk of program failure and may perpetuate a perspective that ITSM doesn't work

Key Strategies:

- Try to lay out the program needs and what needs to be done to be successful as early as possible with this person and gain their agreement before moving forward

- If the above is not successful, consider limiting the scope of the program to what can be done with the resources and organizational leverage that has been provided

- Consider trying to better understand the priorities and goals this person is working under and how the ITSM program may fit into those objectives

- Focus on short term early win efforts as a means for building confidence that time invested in the program is well worth it

- Stress frequent communications on program successes and progress

- If time is an issue with this person, consider "time boxing" with a fixed meeting date and time (e.g. "let's reserve Fridays from 1-1:30pm to review progress")

The Dictator

Characteristics:

- Just wants to get the tool implemented right away

- Feels that process can be easily followed once documented

- Doesn't want to waste time on organizational change, meetings and other activities as it is the job of the staff to "do what they are told"

- Publicly announces that "we'll fire a few heads so they'll know we're serious"

Outcomes if left alone:

- This approach will work for about 6-8 months at which point there may be mutiny or intense pushback

- Staff will have a very negative view of ITSM since it was forced upon them

- May result in higher levels of staff turnover

- ITSM Program typically dies when executive leaves

Key Strategies:

- Try to lay out the program needs and what needs to be done to be successful as early as possible with this person and gain their agreement before moving forward

- See if this person is open to allowing some form of organizational change activities to take place and leave the responsibilities for acceptance with others in the organization

- Focus on early win efforts as a means for softening this person's approach

- If the above is not successful, consider seeking alternative sponsorship for the ITSM program as this approach has never succeeded – this person can do more damage that creates permanent resistance

The Purist

Characteristics:

- Implements ITSM strictly by the book

- Typically addresses issues with "the book says…"

- Tends to derail meeting agendas and objectives with deep discussions on ITSM theory

- May operate on theology versus what is really needed (e.g. nothing else can be done until you get Configuration Management implemented, you should always start with Incident Management)

- Has a hard time with the 80/20 rule – the process must be perfect before you can proceed to the next step

Outcomes if left alone:

- Will slow efforts down taking up tremendous amount of time in meetings on ITSM theory issues

- Generally not a long term team player – may get frustrated and not participate in the effort after a while because the solutions are "not perfect"

- Generates a lot of documentation that is not used by others

- ITSM viewed as a "non-starter" with support staff who feel that the purist approach does not address their day-to-day job issues

Key Strategies:

- Consider placing this person in a subject matter expert role versus leading the effort

- In meetings, monitor for deep discussions that do not move the ball forward, halting those discussions, and then seeing if there is group agreement to schedule them for a later time

- Try to keep them focused through use case situations and discussions (e.g. "walk us through how we handle incident xyz step by step…" versus presenting a generic incident management process flow)

- Recognize that this person could at least provide an outside perspective that the team hadn't considered before (but don't dwell on it if it truly provides no value)

The People Soother

Characteristics:

- Very sensitive to how people are feeling

- Big on organizational change – people are everything

- Tends to avoid conflict situations

- May not make or support decisions until everyone else gets on board

- Uses approaches geared to loose planning with "conflict-free" activities

Outcomes if left alone:

- Will slow down program implementation over conflicts and disagreements that some team members may have

- Implements solutions that only half work because conflicts were not well addressed in the solution

- Gets team resources made up of IT staff that no one else wants

- ITSM team viewed as a separate effort not that is not part of what "real" IT people do or need

- Lasts about 8-10 months or longer until management asks for results at which point the plug is pulled on the program

Key Strategies:

- May be a good resource for the Organization Change team but not in a leadership role

- Pair with a project manage or other leadership resource that is good at managing conflict and getting things done

- Use as a good consultative resource for monitoring acceptance and identifying how well others are adopting ITSM solutions

The Optimist

Characteristics:

- Always displays a "can-do" attitude

- Avoids discussions about issues and challenges that may impeded progress

- Tends to be focused on the tasks at hand versus the larger picture of what needs to be accomplished

- Usually low balls estimates on when things can be finished (e.g. "we can get that tool in by next week", "we can get the process done in a few days", or "should be easy to fix that by Friday")

Outcomes if left alone:

- Missed deadlines and targets that create unplanned team surprises and delays

- May leave solution issues and challenges unaddressed and allow them to fester where they may surface in a big way at a later time

- May also leave the team blind as to issues that already exist that will prevent deployment of the ITSM solution

- Contributes to scope creep that might derail the entire ITSM effort

- Lack of confidence and trust in the team's ability to get anything done

Key Strategies:

- Do not put in a leadership role as this person can be very dangerous – always promising and never getting anything done

- Avoid having this person on the team if at all possible

- If the person must be on the team, make sure tight controls and oversight are provided to ensure expectations are properly set and deliverables are completed when promised

- Do not allow this person to communicate with others outside the program team

The Ostrich

Characteristics:

- Tries to implement the ITSM Program at middle management levels

- Focuses efforts with line staff hoping to bring senior management on board later

- Focuses on "safe" activities: training, certification, incident management, or service desk

- Won't talk to the business

Outcomes if left alone:

- Program may die at the next budget cycle

- Slow progress as other priorities will take precedence

- ITSM Program typically ends up disconnected from the rest of the IT organization

Key Strategies:

- Stick to early wins and small scope solutions that can be done within the management levels and resources that are cooperating with the ITSM Program

- Ensure solutions being implemented are truly addressing business and IT issues and problems

- Continually take efforts to seek senior executive level sponsorship

- Keep communications ongoing about the full benefits of a comprehensive ITSM Program and that current efforts are only accomplishing a small part of those

The Skeptic

Characteristics:

- Promotes the status quo even they agree that things are not always working

- Cites that "ITSM will never work in this organization"

- May feel that ITSM is just a fad and will go away if they wait long enough

- May cite that the IT organization is different from other companies and therefore cannot or shouldn't implement an ITSM Program

- May disagree with ITSM as a viable solution

- Typically starts meetings and conversations with "why this will never work…"

Outcomes if left alone:

- Program implementation may take much longer

- Sows seeds of discontent with proposed ITSM solutions if allowed to fester

- Shuts down positive discussion and progress for solving IT issues and challenges missing potential opportunities that could have worked

- May contribute to low morale for the ITSM team

Key Strategies:

- Focus on program goals and objectives as soon as possible (e.g. "why are we doing this?" What do we hope to achieve?")

- Look for "defining moments" that make ITSM a necessity (e.g. "we just experienced a major outage yesterday – this cannot keep continuing…")

- Consider an approach where an initial objection is then addressed with a solution that specifically targets it – some skeptics will reverse opinion if they see a way out of the issue

- Consider working with the person to change their focus towards what can be done first, coming up with 2 or 3 things that can be done and then only abandoning them if there is no other recourse (sometimes skepticism is a working deficiency that they have gotten used to)

- If any of the above doesn't work, focus on strategies that neutralize the communications and influence of this person (e.g. only include them in selected meetings, make sure they are not viewed as part of the team)

- Recognize that sometimes skeptics have some valid opinions as to why they feel that way – make sure the ITSM Program team has a chance to understand these as ignoring them can create issues down the road

- Recognize that many times positive feedback and head nodding may occur with this person in public but privately they are still skeptical things will work

The Politician

Characteristics:

- Has a specific agenda that is sole outcome of their focus

- Their agenda may or may not coincide with goals of the ITSM program

- Has a well-bred innate understanding of the political landscape of the IT and business organization

- Makes decisions based on personal goals versus what may be best for the program

Outcomes if left alone:

- Can be very positive but only if the agenda aligns with the goals of the program

- Can be extremely negative if the agenda aligns with something else other than the program

- Risk that the agenda might corrupt the ITSM solutions being put into place (e.g. ITSM team targeted incident reduction and better service, but this person only wanted to consolidate the service desk and reduce service desk staff headcount in order to gain a promotion)

- May disappear or unhook from the ITSM Program if political winds about it turn negative

Key Strategies:

- Ensure ITSM Program goals and mission are well communicated and agreed with this person at the earliest time possible

- If agreement can be reached, make sure that there will not be any conflicts with the program ("Let's agree on what we will accomplish...")

- Foster a good working relationship – this person may be very helpful in clearing organizational obstacles and landmines for your program

- Find ways to give this person political credit throughout communication campaigns and events (e.g. "John Smith has sponsored a program that will provide all of us with...")

- Monitor for covert activities – sometimes this person is saying one thing publicly and another thing privately

The Retiree

Characteristics:

- Views their career and program participation as a day job to earn a paycheck

- Counting the days towards retirement (or maybe a different job)

- Generally agreeable in meetings unless it involves overtime work

- Not always willing to take ownership or responsibility

Outcomes if left alone:

- Program may experience much in terms of delays and procrastination towards getting things done

- May experience frequent staff turnover on the ITSM Team

- May experience sudden team shortages as these people take the first opportunity to retire or go elsewhere as soon as they can

Key Strategies:

- Avoid people being put on your team solely as a stopping point for retirement or leaving the company

- If they are on the team, find specific tasks that they can work on such that those tasks can be picked up by others if they leave

- Do not put this kind of person in a leadership role on the team

Identifying WIIFM - What In It for Me?

Identification of the WIIFM (What's in it for me?) for each stakeholder is key towards gaining their buy-in and acceptance. For a variety of reasons, each stakeholder has their own desires, wants and needs that they operate with. It is important to understand what these are and how they may relate to the ITSM Vision and Transformation Program overall.

Addressing these accurately with each Stakeholder will result quite favorably when it comes to overall acceptance of the ITSM Vision and willingness to accept the results of the Transformation Program. The more that everyone can see a stake in the Vision, the more successful you will be at leading your organization towards it. Likewise, those that do not see a stake in the vision will soon move onto other priorities and may even work against you if they feel the Vision runs counter to their own objectives and needs.

Determining the WIIFM for each stakeholder will be done mostly by meeting with them individually and flushing out their concerns and interests. By identifying the type of stakeholder you can get a start towards what some of these might be.

The table below provides some guidance towards this. It lists examples of stakeholder types and identifies possible wants and needs that typically inherent with them:

Type	Typical Wants and Needs
Customers	• Value for money spent if charged for IT services • Better service – can do work faster and more efficiently • More attention from IT • More influence over IT plans and activities
Senior Management	• Meeting corporate quality and improvement initiatives • Improving service • Reducing costs • Delivering services that don't create security risks • Getting ready for a major business initiative or change such as a merger or acquisition • Preventing calls from other business units about IT service issues • Demonstrating compliance with regulatory requirements • Understanding of costs for providing IT services
Middle Management	• Meeting business unit targets and objectives • Keeping staff and direct reports motivated • Preparing for a major business initiative or change • Preventing calls from upper management about IT service issues • Keeping IT costs in line with budget targets and objectives
IT Delivery and Support Staff	• Recognition for work performed • Eliminate firefighting • Working smarter instead of harder and longer • Reducing time spent on non-value work
IT Technical Architecture Staff	• Major support for IT Architecture standards and initiatives • Major support for IT governance • Recognition for technical solutions that support desired processes and business objectives • Increased visibility for architecture role and activities
Sponsors	• Return on investments made in ITSM • Successful implementation program • Meeting program objectives

Type	Typical Wants and Needs
Users	• Get work done without service disruptions • Get work done without spending time dealing with IT technical issues and problems • Better communications about service outages and root causes for them
Customers	• Receiving service value for money • Ensuring they have made the right selection in providers for their services • Ensuring that business processes are well supported
Vendor Suppliers	• Better relationship with your business organization • Opportunity to increase services or length of service contracts • Clear definition of service roles and responsibilities • Clear roles and responsibilities for handling problems and incidents to reduce finger pointing and blame
IT Development Staff	• More efficient process for placing IT solutions into production • Increased business satisfaction with new IT solutions • Clear roles and responsibilities for handling problems and incidents to reduce finger pointing and blame between development and operations • Ensure deployment and operational considerations baked in as solutions are developed versus scrambling at the end before deployment
Change Agents	• Recognition • Chance to learn something new • Work on interesting projects • Make a positive difference in the organization
Stakeholder Initiatives	• Gain increased support and sponsorship for initiatives already underway • Share like ideas and improve solutions already under design • Increased influence and visibility for the initiative being worked on
Champions	• See desires turn into actions that make a positive difference • Gain visibility and recognition for opinions and foresight • Desires to see things done in the best possible way

External Barriers to Change

While people present some of the greatest challenges to adoption and acceptance, external factors also can exist. These should not be ignored. Sometimes IT organizations will treat them as the big elephant in the room that no one wants to talk about. If any of these exist, make sure they are dealt with directly.

Here are examples of external threats to your success that have been seen in many IT organizations and what you might do about them.

External Threat	Possible Solution
Current IT organization is in technical silos and there is no appetite to change that structure.	Establish a logical or virtual ITSM organization made up from individuals in the different silos. Make sure this has a direct line to IT leadership.
CIO/CFO sponsorship is weak such that the CIO won't make decisions or champion the effort when threatened.	Seek out a strong influencer who has a direct relationship with the CIO who can also take on decisions on the CIOs behalf. If that won't work, consider a strong initial win effort that gets noticed by the business. Sometimes this is enough to get a key leader to stick their neck out a bit.
Time constraints – typically driven by a need to get off an old tool license or consolidate an IT department.	Carefully scope efforts to limit what can be done with the available time. As a backstop, consider extending use of the tool without vendor support or establishing a target organization as early as possible with decision autonomy that then "hires in" the other IT organizations.
Organization won't abandon the status quo.	Put in a strong organizational change strategy. Initially do all the right things in communicating benefits and why things have to change. With this, go after a "tighten the screws" strategy where over time it becomes uncomfortable and ultimately politically dangerous to stay with the status quo.

External Threat	Possible Solution
Sudden business initiatives or other unplanned events force IT to rethink its priorities about doing an ITSM project.	Consider re-scoping the ITSM initiative to focus only on those areas that will assist with the change in direction. For example, if an ITSM effort can't be done because suddenly a major data center move needs to happen, some solutions in the area of capacity planning, event management, service transition and configuration management will be critical towards accomplishing that goal.
Too many organizational politics and emotion in making decisions and establishing priorities threatens to delay or stop the ITSM effort.	Nothing succeeds better with this than presentation of hard facts and numbers to drive the decision making. Include trends with this. For example, if we don't address our high incident rate now, by next year, the rate will have nearly doubled. Keep measuring against that prediction to demonstrate that reality is following the predictions that were set.
Punishment culture in which people are severely blamed for mistakes or other issues such that they don't honestly work to raise issues and fix problems.	Early on, set a practice that is numbers and process driven. Conduct offsite sessions to review results and work with key executives to send a strong message that processes are at fault, not individuals. Consider blind improvement suggestion boxes or other approaches where people can raise issues without feeling threatened.
Ethical issues such as hiding facts, misrepresenting numbers, throwing people under the bus for political gain, cooking the financial books or other misdeeds.	While it may be possible to cleverly participate in similar behavior to further the ITSM initiatives, it may be best to get totally out of this one and go somewhere else.

Signs That Change Is Not Taking Hold

Always be on the lookout for these issues not only during transition to ITSM solutions, but also after those solutions have been put into place:

- The old culture comes back during a major outage, crisis or challenge. "Throw out ITSM for now – we have to get this done…" they will say.

- The ITSM solution culture and behaviors are slowly becoming isolated and not spreading throughout the organization. "That's for the ITSM group", they will say, "They're not really what the rest of us do".

- No ongoing continual service improvement activities are taking place. Essentially saying "…the ITSM project is done…therefore we must have finished improving…"

- Service reports, scorecards and dashboards are no longer being used by management and findings in those reports are not being acted upon.

- Other IT priorities have totally replaced work on ITSM initiatives.

- Changes in IT management where the new management focuses elsewhere and de-prioritizes the ITSM effort.

- Service Owners, CAB members, Managers or other support staff no longer attend key ITSM meetings.

- IT outcomes such as incidents and service breaches are getting worse instead of better.

Should you start to see any of the above, it may be time to redouble your organizational change efforts to effect a turn around. It is not unusual for people to slip back into their old behaviors unless continual reinforcement is in place.

Chapter 7

Building Communications Step By Step

Overview of the Approach

You can build and architect a great IT Service Management solution, but it will ultimately fail if people refuse to use it. This refusal comes from many directions. Examples include:

- Inability to understand how new tools and processes work

- Fear that new tools and processes may expose personal weaknesses and flaws

- Feeling that new tools and processes will get in the way of completing day-to-day work tasks

- Perception that the effort is not as important as other things that need to get done

- Disagreement about the approach that solutions are using

- Personal bias towards specific tools and processes that did not become part of the solution

- Fears that personal influence or job importance may lessen once the new solution is in place

- Concerns that the quality of IT support and delivery will lessen once the new solution is in place

- Fears about loss of control from middle management and supervisors

The workforce needs to understand the need for and the impacts of the changes taking place. That it possesses the capabilities and motivation to change. The key objective of the communications approach presented here is to prepare the workforce for working with new or changed processes and technologies.

Communications activities need to be integrated with the project team activities. It is strongly recommended that a communications track be added to the project plans that will run in parallel with other implementation and transition activities. If left to the end of the project or ignored, there is great risk that the project will get delayed or fail entirely.

For that reason, the approach presented here mirrors very closely with the ITSM Service Lifecycle or project waterfall approach. If the team is taking a SCRUM or Agile approach to their efforts, then the tasks presented here also need to be aligned with the output of those efforts. You cannot change the way people work and interact simply by throwing new technology at them. It is important to highlight that these activities must be highly integrated with the project team implementation activities, following the same phases as the project team phases.

Sustainable acceptance of change occurs when people understand how to use the new technologies, how to use the new processes, and agree with the vision behind the changes. The mission of the communications approach presented here is to confirm that the impacted workforce clearly understands the need for and the impacts of the change, and that it possesses the capabilities and motivation to change. The activities described will be used to accelerate the impacted workforce's adoption of the changes and reduce the risk of disruptions and confusion too often experienced when these changes are first put into place.

Key aspects of the approach presented include:

Challenge	Approach Taken
Adoption and Learning	Conduct of a series of communication and learning events that repetitively introduce stakeholders to new skills and ways of working with the new solution
Managing Resistance	Allow stakeholders time to get comfortable with new solution through hands-on use of tool, case examples, walkthroughs and allow them opportunity to raise issues and suggestions
Transition Readiness	Test stakeholders on skills being learned and take them through the transition approach with a final transition walkthrough prior to going live
Managing Stakeholders	Maintain an inventory of stakeholder types and target selected knowledge and training for each type – monitor stakeholder progress through each learning campaign to validate readiness for transition

The suggested methodology for communications activities is as follows:

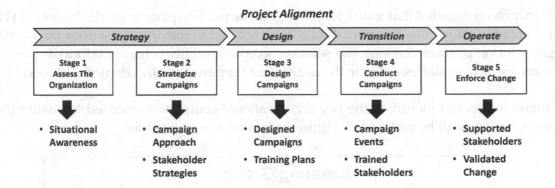

The effort is broken down into 5 key stages. Each stage aligns with a typical project phase. It is highly recommended that the activities within each stage are embedded into each project phase as it takes place. This will avoid delays to the overall project. Each of these stages are described in more detail in succeeding chapters.

Stage 1 – Assess the Organization

This stage is about gaining awareness of the current IT organization and the challenges that lie ahead. Key activities include assessing the organization and creating the outputs of that assessment that will help shape and strategize the campaigns and approaches you will use to effect changes.

Stage 2 – Strategize Campaigns

In this stage, and using the outputs of the assessment, stakeholders will be identified along with their information needs. A high level communications plan will be developed that identifies the communications and learning events that will take place along with key timeframes and milestones.

Stage 3 – Design Campaigns

In this stage, each campaign event identified earlier will now be fleshed out in detail. This includes design for workshop and training events, design of the materials to be used at those events, meeting agendas and how progress will be tracked and monitored.

Stage 4 – Conduct Campaigns

In this stage, communication and training events are carried out.

Stage 5 – Enforce Change

In this stage, activities are undertaken to reinforce the change that has taken place. Stakeholders are supported as needed to get comfortable with the change. A final validation is done to ensure change has taken place successfully.

Organizing For ITSM Transition

It is highly recommended that a structured Program be put into place to guide the overall ITIL Implementation effort. The Program should be structured to maximize the need for solution design, solution communications and solution acceptance. This is highly critical as process solutions are only as successful as to the degree that they are actually adopted and used.

The following model identifies the key organizational components needed to ensure that process solutions will be embedded within the business organization:

Steering Group

Program Office

Program Governance and Measurement

ITSM Advisory Team

ITSM Extended Team

Organizational Change Team	ITSM Core Teams	ITSM Technology Teams

Subject Matter Experts

Stakeholder Initiatives Teams

Vendor Suppliers

Steering Group

Sets project direction, makes key decisions and provides funding and final approval of program deliverables. Represents the ITSM Implementation Program to other business units. Key activities are:

- Champions process solutions across the business enterprise
- Conducts periodic meetings for Program progress and issues
- Provides final review and approval of program deliverables
- Coordinates approvals from non-IT business units as necessary
- Identifies and appoints program team members
- Coordinates major program decisions that are escalated to steering group on a timely basis to meet program objectives

Program Office

Oversees the day-to-day management of the Program efforts. Ensures correct and timely deliverables completed and sent to the Steering Group and ensures the overall objectives of the Program are met. Accountable for cross project coordination to make sure that program goals are being met.

Key activities can include:

- Reviewing project plans from other teams for completeness
- Integrating plans from other teams into a comprehensive solution plan
- Monitoring and reporting progress of project plans and activities
- Providing other teams with standards for reporting progress and documenting their plans and activities
- Gathering and collating status report information
- Administering Program document repositories and web pages
- Collecting labor hour information on program participants
- Managing program email distribution lists
- Managing and publishing the program calendar
- Setting up program meetings and schedules
- Coordinating travel arrangements for program participants

Program Governance and Measurement

Provides the Program Manager who controls and manages overall scope of the program effort. Key activities include:

- Acting as an escalation point for program changes

- Coordinating research to identify impacts of proposed program changes

- Coordinating approvals for program changes

- Acting as a focal point for communicating new technology/process ideas to teams

- Utilizing a project change management process to underpin scope containment

- Tracking and reporting on overall program metrics

- Responsible for the overall project objectives.

- Provides direction to the project teams for deliverables due as well as the overall status of the project.

- Co-ordinates with other project managers.

- Provides status of work in progress and/or issues to the Executive Steering Committee

- Develops project work plan, schedule and staffing requirements.

- Communicates as required to executive management or IS&TS staff.

- Conducts weekly change, issues and status meetings to track progress and risks.

- Ensures that outstanding project management, process implementation and design requirements and/or issues are being addressed.

- Communicates activities and status of the project throughout their working environment.

- Schedules workshops and meetings as required.

- Provides overall leadership and management of the project.

ITSM Advisory Team

Advisory Stakeholders have limited participation and do not work day-to-day with the Program effort. They are typically high level managers and directors that need to have awareness of what the Program is producing and may be consulted for key decisions. These stakeholders:

- May provide input/review of key program deliverables

- May provide direction and guidance on program decisions and issues

- May assign others within the department represented to work on teams

- May be involved only on a need-to-know basis

- Reviews output of the implementation effort

- Provides key decisions and approvals on a timely basis to meet implementation project needs

- Assigns other department personnel to serve as additional Advisor and Extended team members as needed

- Works in conjunction with other Advisor or Extended Team members within the department as needed

ITSM Extended Team

Extended teams are the key vehicle for obtaining overall business buy-in and agreement to the solutions being developed. They leverage communications between ITSM core implementation teams and other business units by providing feedback, acceptance and input from those units.

Each team member is linked with one or more process core teams.

Those designated as Extended Team members actively participate in the program effort part time on a day-to-day or week-to-week basis to produce or assist on program deliverables. These stakeholders:

- Actively participate in the development of ITSM solutions

- Represent the particular needs of a business or IT unit

- Provide input to the Core Teams from the unit they represent

- Communicate solutions/issues to the unit they represent

- Obtains buy-in from the units they represent

Use of the Extended Team is to leverage communications up and back between the Core Teams and the rest of the business enterprise. It is the Extended Team that has the responsibility and accountability to make sure that the business units they represent agree and will use ITSM solutions. This allows the Core Teams to focus on building ITSM solutions without getting bogged down in myriads of meetings and communications,

Organizational Change Team

Manages and coordinates tasks related to developing and leading the organizational change effort to get people to adopt, use and align to the ITSM solutions being implemented. Essentially carries out communications activities as described in this book. A summary of those activities includes:

- Performs stakeholder management activities to identify stakeholder concerns and issues with solutions being developed

- Monitors stakeholder acceptance/rejection of solutions being developed

- Crafts and controls key communications and messages about the implementation effort

- Identifies opportunities to win acceptance of solutions being developed by those who are impacted

- Identifies channels for communications and builds the overall communications plan

- Develops a Resistance Management Plan to provide strategies for dealing with rejection or resistance to solutions being developed

- Ensures appropriate levels of the organization are involved and demonstrating active commitment and leadership to the solutions being developed

- Develops, conducts and administers training to provide impacted stakeholders with the skills needed to operate ITSM solutions successfully

- Designs and builds the communication campaigns and events

- Provides facilitation for key meetings and workshops as needed

- Communicates with users of the process as to what is expected of them

- Leads meetings and working sessions in a neutral manner to ensure goals and outcomes of those sessions are being met

- Develops session detail agendas and agrees these with those involved

- Develops discussion strategies and methods to ensure all participants are involved and to obtain consensus on key decisions in an efficient manner

- Monitors sessions to make sure all sides of discussed issues are being considered and that session groups do not "go with the flow" unless truly in agreement

- Trains or coordinates training for the users of the solutions on tools and procedures

- Prepares training material if required

- Provides input into the development of the training material if required

- Is aware of the project plan and highlight issues as they arise to the Program Office

ITSM Core Teams

This consists of a number of implementation teams usually split by processes or groups of processes. These are the teams that get the work done from a process perspective. They perform the actual implementation work. They coordinate solutions and activities with the Technology teams and Extended Team members. They lead presentations and demonstrations of ITSM solutions as they develop.

Key activities include:

- Develops project work plan, schedule and staffing requirements for each process solution area

- Assesses the current state of readiness and effort required to implement the processes, tools and organization

- Designs and builds processes for each ITSM solution being developed

- Coaches the users of the process on tools and procedures

- Communicates with the Process Owner on process design, status and issues

- Manages resources during detailed design and implementation

- Ensures the solution documentation is maintained.

- Participates with the Organizational Change Team at communication events

- Provides tooling requirements and manages changes to tools and organization to support the process as required

- Ensures interfaces to other processes are working and coordinated

ITSM Technology Teams

This consists of a number of implementation teams usually split by processes or groups of processes as well as technology components and platforms. These teams get the work done from a tooling perspective. They implement ITSM tools, operate those tools and integrate them with the current operating environment. The customize tools based on requirements as provided by the Core Teams.

Key activities include:

- Ensures the tool architecture meets the strategic needs of the IT organization

- Co-ordinates product selections

- Designs and implements ITSM tooling solutions

- Customizes tools to match solution needs and requirements

- Ensures maximum integration of tools

- Coordinates technical resources to optimize tooling implementation and customization efforts

- Communicates tool architecture to program teams

- Interfaces to vendors as needed

- Provides input to the detailed design for the processes

- Customizes the tools based on the detailed design

- Tests the tools

- Assists in developing procedures to install the tools

- Sets up the education environment to support training needs

- Assists in development of ITSM transition plans

- Assists in conduct of ITSM deployment activities

- Resolves problems with the tools as needed

ITSM Coalition Teams

If you have a very large IT organization with multiple locations and geographies that will all be impacted by the ITSM program, then you may consider use of Coalition Teams. Coalition teams act like Extended Team members in that they represent one or more IT sites that will be using the solutions developed by the program. Members may represent one site or a group of sites.

Members are responsible for ensures that solutions are rolled out to the organizations and business units they represent. They also provide input to the Core Team on solutions as those solutions are developed.

An example of this might be a Data Center Operations Manager who represents other Data Center Operations Managers at other processing sites. This avoids overly large stakeholder teams. For example, if there were 196 processing locations, you might leverage all of them with a much smaller number of representatives. Each representative might have 10-30 locations they are responsible for. In this way, you avoid having an extended team with 196 stakeholders actively involved. It is a leveraging strategy designed to quickly deploy across many locations and geographies.

Key activities of this team can include:

- Serving as in an Extended Team kind of role for others in similar organizations and business units

- Deploying solutions to the organizations and business units they represent

- Meeting on a periodic basis with the organizations and business units being represented to communicate Core Team decisions and program progress

- Feeding back information to the Core Team from organizations and business units being represented on concerns and issues that may exist from those entities

- Conducting implementation activities that are local to the locations representative

- Conducting or coordinating training local to the locations represented

- Gathering data, configurations or other information needed by Core and Technology teams to develop their solutions

- Assisting in analysis and resolution of process and technology issues that may arise local to the organizations they represent

- Assisting with procurement and contracting activities for local needs

Subject Matter Experts

Provides expertise in technical, operational and managerial aspects for the design and implementation. Not a full time team member but participates in the implementation as required. Provides specialized expertise in the design and implementation of the ITSM solutions. Key activities are:

- Provides technical, operational, business and/or managerial subject matter expertise

- Provides input into the design of the procedures, tools or organization as required

- Develops solutions as required

- Supports the development and execution of test scenarios designed to validate the functionality of the design

- Validates designs for processes, tools and organization and provides recommendations

- Communicates activities and status of the project throughout their working environment

- Provides consultative and facilitation support to the other teams

- Assists in creation of the project plan

- Provides Intellectual Capital as required during the Implementation Project, based on external experiences

- Coaches team members as required.

Stakeholder Initiatives Teams

It is possible that other related projects and programs within the company are already being undertaken or considered that relate or have impact to the ITSM program. Examples of these might be a Change Management initiative occurring in a different business unit or a Six Sigma improvement effort that the corporate office has been running.

For each initiative identified, the Core/Extended/Advisor teams will need to determine a strategy for how to engage. Examples of these might be:

- Attend regular meetings held by the initiative

- Invite initiative teams to program working sessions and events

- Engage on a need-to-know only basis

- Make an initiative team member an Extended Team member

- Don't engage at all

Vendor Suppliers

This represents 3rd party suppliers to the ITSM effort. Supplier representatives can participate as Extended Team members. Consulting service providers may also participate on ITSM Core Teams. It is strongly recommended that each supplier be represented. If it does not make sense to have suppliers actively participate in any of the teams, then appoint a representative to each supplier that is already part of the Core or Extended teams.

Program Structure Example

The organization structure below represents one possible way to organize your program using the model described earlier:

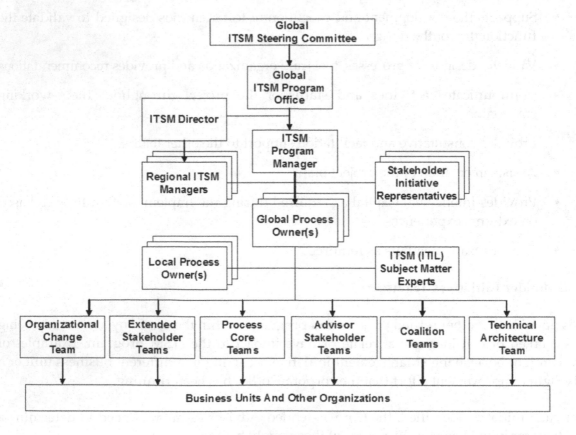

In the above organizational structure, each organizational unit conforms to the model presented earlier. The management units make up the overall model in the following ways:

Organizational Unit	Description
Global ITSM Steering Committee	Consists of ITSM Steering Group members who will guide and oversee the direction of the entire implementation effort.
Global ITSM Program Office	Consists of an ITSM Program Manager and Administrative Analysts who will oversee and manage the entire implementation program on a day-to-day basis.
ITSM Director	In this example, there is a new IT Service Management organization that was started within the company. This shows where the Director of that organization links to the overall program.
Regional ITSM Managers	Since this is a global organization, regional ITSM Managers have been appointed to represent and oversee Program activities within key geographic areas of the company (i.e. Americas, Asia and Europe/Africa).
ITSM Program Manager	This calls out the Program Manager role that is part of the Program Office. This person also is part of the new ITSM Director's organization in the example.
Stakeholder Initiative Representatives	In the example, it was discovered that several IT service improvement initiatives plus a corporate TQM initiative were already underway. Selected representatives from these efforts have been linked into the Program.
Global Process Owner(s)	Process ownership at the global level has been established here. These owners will lead and coordinate their process efforts from a global perspective.
Local Process Owner(s)	Process ownership has also been established at the regional levels. These groups report to a regional ITSM Manager, but work directly with and for the Global Process Owners as part of the Program.
ITSM Subject Matter Experts	This unit covers Program experts and consultants that will be assisting and helping with many of the ITSM Implementation Program tasks.
Organizational Change Team	Represents the team that will focus on business cultural change and awareness to meet the Program objectives.

Organizational Unit	Description
Extended Stakeholder Teams	Represents Extended Team Stakeholders that will assist the Process Core Team efforts on behalf of their business units.
Process Core Teams	Represents Process Core Team (See Program Role Descriptions) members, each of which are assigned to one or more processes.
Advisor Stakeholder Teams	Represents Advisor Stakeholders that have a major interest in the Program outcome. This group may be asked to help make key decisions on solutions chosen and to identify other Stakeholders that should be involved with the Program.
Coalition Teams	Consists of groups of people that represent major geographical or large groups of organizational units within the company. An example might be a selected group of people who will represent all the data processing centers in the company. This group will be actively involved in ensuring process solutions developed globally can actually be deployed within the units they represent.
Technical Architecture Team	Consists of personnel who will focus on ITSM technical architecture and tool solutions.

Considerations for Choosing Team Members

Here is a list of general considerations when choosing team members to serve on each team:

- Consider all stakeholders impacted by the solution and look for a cross section of functions, roles, and management levels that truly cover the business organization

- Look for a mix of change leaders and adapters

- Include some key resistors to challenge the approach and concepts

- Include people who want to participate as much as possible, even if only a small role

- Encourage those who resist but whose input is critical to the transformation effort

- If it appears management is slow in assigning team members, ask for an interim member who can serve until they make up their mind

- Don't let politics get in the way with the selection as the recommended team structure provides ample opportunity for all to get their say

- Be careful about management assigning just anyone because they happen to be available as this will create rework when the solution progresses and they determine it was the wrong person

- For teams that are staffed with senior leadership, it is okay to have them use representatives in their stead as long as they agree that the person chosen truly represents their interests

- Try as much as possible to get some business representation on the teams where it makes sense as input from those sources ensures the ultimate solution will provide value to the business

- If team members are constantly busy, consider "time boxing" them into a fixed number of hours each day or week (e.g. "Friday from 1-3pm is just for us")

Team Resource Requirements

You are almost sure to be asked at the start of the effort how much time is needed from team members to participate. While this can vary depending on the size of the IT organization, the company or how fast or slow you wish to proceed, here are some general guidelines:

Team Role	Commitment Guidelines
Steering Group	Up to 4 hours each month.
Program Office	Full time for a program manager for large efforts, up to half time for moderate efforts.
Program Governance and Measurement	Full time for a program manager for large efforts, up to half time for moderate efforts.
ITSM Advisory Team	As a general rule, about 4 hours each month. Some Advisors may want to participate more, others not at all but want progress reports or other information. For this team, use the 4 hour commitment but let them decide how they want to participate.
ITSM Extended Team	About 4 hours or a half-day each week of the effort. This time is for reviewing solutions being built, addressing issues, working with the units they represent, or providing input to other ITSM teams.
Organizational Change Team	This is a full time role for a moderate to large effort. Smaller efforts can probably be half-time.
ITSM Core Teams	This is a full time role for a moderate to large effort. Smaller efforts can probably be half-time.

Team Role	Commitment Guidelines
ITSM Technology Teams	This is a full time role for a moderate to large effort. Smaller efforts can probably be half-time. While there is a tendency to think this role is done after the tool is installed, reality has shown that there is still work to be done for testing, addressing tool issues, and ongoing maintenance.
ITSM Coalition Teams	About 4 hours or a half-day each week of the effort. While similar to the Extended Team, members can also end up participating up to full time if there is a lot of implementation work that needs to be done locally.
Subject Matter Experts	Generally half to full time in short durations to address specific issues. If the work is needed longer term, then consider moving members to the Core or Extended Teams.
Stakeholder Initiatives Teams	This will vary by the strategy chosen for working with each initiative. If it was decided to interact fully, then it is a half to full time role. If only to exchange information, then about 4 hours each month.
Vendor Suppliers	Generally half to full time in short durations to address specific issues. If the work is needed longer term, then consider moving members to the Core or Extended Teams.

Chapter
8

Stage 1 - Awareness

In this stage, the key objectives will be to identify who your stakeholders are that will be impacted by the change and understand the culture of your IT organization. The former is key in that communication events and eventual adoption of the solution will come from those stakeholders. The latter is also important as information about your culture will be used to help identify what activities will be needed for successful adoption.

This stage spends a lot of attention on understanding stakeholders. It is key to clearly understand what they will react to. Many might be in agreement with you, nod their head, and give you positive feedback, but doesn't mean they will follow through or fund your efforts. Leadership might agree, but doesn't mean they will show up at meetings, champion or walk the talk. For this reason, it is important to understand who will be impacted by your ITSM solutions and what will propel take positive action.

Approach Overview

For this stage, the following activities should take place:

Step	Action
01	Conduct organizational awareness
02	Identify stakeholder types
03	Create a Stakeholder Map
04	Identify stakeholder priorities
05	Identify stakeholder current acceptance level
06	Identify stakeholder influence
07	Identify stakeholder wants and needs
08	Define high level communications programs

Key outputs from this stage will be:

Situational Awareness Findings

A summary of key organizational issues, cultural weaknesses and behavioral challenges that will need to be considered as part of the overall communications strategy.

Stakeholder Map

Listing of key stakeholders, their level of influence, wants and needs.

Communications Plan Outline

General outline for the Communications Plan. For now, some sections describing the overall ITSM Program vision, stakeholders and document purpose can be built. Details on campaigns and events will be fleshed out in more detail in later work stages.

Conducting Organizational Awareness

Let's start by looking at the overall company. Most companies go through a continuous cycle that looks like the following:

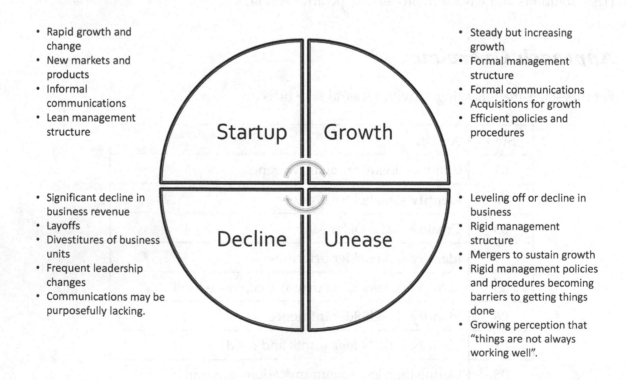

- Rapid growth and change
- New markets and products
- Informal communications
- Lean management structure

- Steady but increasing growth
- Formal management structure
- Formal communications
- Acquisitions for growth
- Efficient policies and procedures

Startup | Growth

Decline | Unease

- Significant decline in business revenue
- Layoffs
- Divestitures of business units
- Frequent leadership changes
- Communications may be purposefully lacking.

- Leveling off or decline in business
- Rigid management structure
- Mergers to sustain growth
- Rigid management policies and procedures becoming barriers to getting things done
- Growing perception that "things are not always working well".

Where is your company on this scale? Where you sit may indicate some of the challenges you have. For example:

- **Startup** will present challenges in getting people's time and commitment to change. Too much is going on. Focus here will probably be on risks. Can the company succeed without ITSM? Is ITSM needed to reverse a decline or save money? Are customers or regulators demanding proof of ITSM practices?

- **Growth** may be the best place for companies to be, but they are also the most resistant to change. After all, things are growing and getting better. Why change now? Focus here on issues around scale. Can the company continue to grow without ITSM in place? Do regulatory issues or operating requirements threaten the ability to continue to grow?

- **Unease** occurs as the company's growth is starting to slow or stagnate. There is a feeling that some things are not working well. In this environment, there may be pockets of forward thinking people who are willing to change even if the rest of the company is not yet ready. Efforts should seek out and focus on these stakeholders. Can ITSM practices reduce operating risks and help the company grow or reverse course?

- **Decline** may present the greatest acceptance of change since there is more desperation to operate differently. What can be done to reduce costs? Gain new markets? Reverse the decline? This stage requires focus around hard benefits that can be seen. Promise of soft benefits (e.g. better handling of incidents, less outages, etc.) does not play well in this stage.

The messaging and frequency of communications will need to be planned for depending on what stage your company is in.

Now let's look at your IT organization. Which best describes the general culture?

Culture	Characteristics	Suggested Strategy
Predatory	• Few rules • No shared vision • Individual rules • Exploits employees and customers	ITSM initiative should be abandoned
Frozen	• Aversion to risk taking • No one acts unless they have permission • Management wants to maintain status quo • Can lose job if stepping out of chain of command	Sponsorship at the highest levels needs to be in place with continual and very frequent communications
Chaotic	• Acts unpredictably • Can't focus • Can't walk the talk • Lack of common focus across divisions • High turnover • Frequent reorganizations • Can't marshal support to act on change initiatives	Pull together one or more teams, focus on highly publicized short term wins to gain attention and tilt other groups towards ITSM
Political	• Internal jockeying for influence dominates management agenda • Competing factions, practices are in place • ITSM at war with other ideas • Whole never equals value of its parts • Hero building takes precedence • One portion of the whole achieves dominance over whole organization	Focus on individual stakeholder acceptance of ITSM including them frequently in decisions making sure all stakeholders have been reached and feel they have contributed to the solution

Culture	Characteristics	Suggested Strategy
Bureaucratic	• Customer needs fall below those of IT • Priority is on checking, monitoring, oversight and strict financial controls • Jobs structured around layers of the above activities • Innovation and problem solving are smothered • Mission may be outdated but will work to justify it anyway • Behavior molded around processes and procedures versus people and productivity	Strong sponsorship should be obtained not only by executive management but also key middle managers and influencers with very frequent communication events taking place
Service	• Customer is priority • Focus on customer needs and requirements • Focus on customer satisfaction • Continual service improvement is in place • People empowered to serve customer	Voice of the customer will be key when choosing solutions and it will be important to make sure all key stakeholders are involved in the decision making
Innovative	• Product innovation focus • Customer service focus • Workforce effectiveness • Exotic new technologies • Lack of hierarchy may be a challenge • Little structure or job titles • Anticipates or creates change • Low turnover • High levels of customer and employee loyalty	ITSM may be seen as a barrier or unnecessary overhead to getting things done unless tied to innovative ideas – seek out a highly influential stakeholder that is well respected by the line organization to help influence them towards the solution

The Communications Tools and Techniques chapter of this book also offers a number of other analysis tools to assess the overall culture and where labor is being spent for the current IT organization. The output of these tools will provide some direction as to what strategies to use within your communications plan and the kinds of messages that will resonant best with stakeholders.

Identifying Stakeholder Types

Stakeholders are those who will have a stake in the ITSM Transformation program and its results. They also include those who need to change any behaviors and actions as a result of the ITSM solution being developed. It is absolutely critical to understand and manage your stakeholders at all times throughout the lifecycle of the transformation effort. Failure to do can easily result in resistance to implementing changes, program delays or even serve to kill the entire effort.

A Stakeholder Map is key to monitor and manage your stakeholders throughout the life of the transformation effort. In this step, you will be laying the groundwork for who these stakeholders are.

Independent of any individuals in your organization, identify the generic types of stakeholders that will be impacted by your ITSM solution. An example of this list can include any of the following:

- Sponsor
- Customer (one who buys IT services)
- IT Support Staff
- IT Middle Management and Supervisors
- IT Executives
- Call Center Agents
- IT Developers
- IT Architects
- Technical Support Staff
- Operations Support Staff
- Training Support Staff
- 3rd Party Suppliers and Vendors
- Business Unit Users
- Business Unit Managers
- Business Unit Leadership
- External Customers (that directly use IT services)
- Quality Assurance Staff

When you look at the ITSM solutions you are putting into place, think about all the kinds of people that will be impacted by them.

Creating a Stakeholder Map

Now create a Stakeholder Map. This can be done on a simple spreadsheet. As a first step, populate a column in that spreadsheet with the types of stakeholders you have selected. This spreadsheet will continue to be elaborated as you go through this chapter of the book. It will be a key tool to use as you go forward with your communications plans and events.

Next, in another column, place names that map to each of the stakeholder types you identified earlier. The example below illustrates this:

Name	Title	Type
Jim Jones	VP of IT Architecture	Senior Management
John Smith	Manager, Sales Operations	User
Becky Thomas	VP, Sales	Customer
Rachel Marrow	VP, Service Delivery	Sponsor
Ralph Johnson	Business Representative	Customer
Siddrah Kumar	VP, IT Operations	Senior Management

Identifying Stakeholder Priorities

Here you will be identifying the priority of the stakeholder in terms of relevance to accomplishing the ITSM objectives. This is to make sure that the team is spending time appropriately with each stakeholder. For example, you would not want to prioritize time and effort on working with a stakeholder who might have little influence or decision making authority.

The following table identifies some recommended priorities and their criteria:

Priority	Criteria
Immediate	Someone you think is essential to contact immediately because of the critical nature of their support or influence on others. These are the first people you want to contact about your change.
Soon	Someone whom you need to contact soon, but maybe need the help of another stakeholder in order to influence them. Also, you may need to gather additional information before contacting them.
Later	Someone who for any number of reasons does not need to be immediately involved and perhaps it's better if they are not immediately contacted about the change so that you know how to involve them at the right time, in the right way and for the right reasons.
Unknown	Someone for whom you are not quite sure fits into any of the above priorities.

In your Stakeholder Map, create an additional column for the Priority. For each stakeholder, list the priority to be used.

Name	Title	Type	Priority
Jim Jones	...	....	Immediate
John Smith	...	....	Immediate
Becky Thomas	...	....	Soon
Rachel Marrow	...	....	Unknown
Ralph Johnson	...	....	Later
Siddrah Kumar	...	....	Unknown

For the above, columns greyed out didn't change, just done to make room for further columns that we are adding as the Stakeholder Map spreadsheet grows.

Identifying Stakeholder Acceptance Levels

For each stakeholder, identify their current acceptance level. At this point, it is a general guess or estimate based on what you may currently know about the person. You can also seek out others who may have an opinion also.

For each stakeholder estimate their acceptance level. Are they championing? On board? Neutral? Working against you? This identifies your best opinion on how supportive you believe the stakeholder to be towards the ITSM effort. The table below lists some common acceptance levels:

Level	Description
Negative	Will not change their own behavior and they will publicly and actively work against the effort.
Against	Does not believe in the ITSM Program and will not participate in it nor make any changes as a result of it.
Neutral	Is against the ITSM Program but will hold off participation and changes for it until it appears others are moving forward.
Positive	Willing to participate in the ITSM Program and make the necessary changes to meet program objectives.
Champion	Willing to actively promote the ITSM Program and lead others to make the needed changes to meet program objectives.
Unknown	Stance is unknown in relation to the ITSM program. Try to assess further during communication campaign events.

In your Stakeholder Map, create an additional column for the Acceptance Level. For each stakeholder, list the acceptance level to be used.

Name	Title	Type	Priority	Acceptance
Jim Jones	...	....	....	Positive
John Smith	...	....	....	Champion
Becky Thomas	...	....	....	Neutral
Rachel Marrow	...	....	....	Unknown
Ralph Johnson	...	....	....	Neutral
Siddrah Kumar	...	....	....	Against

Identifying Stakeholder Influence

This identifies your best opinion as to how influential the stakeholder is towards the ITSM Program. For example, the stakeholder may be a key decision maker that can get things done. In another example, they may act as an influence to convince others. It is important to accurately recognize where each stakeholder is with this.

Do not let organization titles and hierarchy levels blind you to the real influence role that a stakeholder may have. For example, that overbearing Vice President in one of your business units may actually be making all their decisions based on input from the meek and quiet Assistant Manager that works for them.

The following table identifies some recommended influence roles and their criteria:

Influence Role	Criteria
Decision Maker	Will have significant say over direction and content of the ITSM Vision. Is able to back up decisions with funding and resource commitments.
Influencer	Influences others who are authorized to make key decisions. Will have significant say over direction and content of the ITSM Vision because of influence versus having funding and resource commitment authorization.
Controller	No significant influence or decision making authority, but manages and leads others once convinced about the merits of the ITSM Vision.
Information Provider	Has no influence or decision making authority, but is adept at keeping their ear to the ground and reading the tea leaves (as it were) about how the organization might respond and who else should be consulted. May very adept at politics within the corporation.
Subject Expert	Has critical expertise in a subject area that is impacted by or contributes to the Vision and its approach. Usually consulted for input. Stakeholders in this category may need to be consulted by others on the Vision Team to gain insight on an issue or subject matter.
Participant	Little or no influence, but will be a participant in the ITSM Vision once directed to do so.
Unknown	Influence role is unknown.

In your Stakeholder Map, create an additional column for the influence level. For each stakeholder, list the influence level to be used.

Name	Title	Type	Pri.	Acc.	Influence
Jim Jones	...	....	....		Decision Maker
John Smith	...	....	....		Influencer
Becky Thomas	...	....	....		Influencer
Rachel Marrow	...	....	....		Unknown
Ralph Johnson	...	....	....		Subject Expert
Siddrah Kumar	...	....	....		Controller

Identifying Stakeholder Wants and Needs

Identification of the needs of each stakeholder (e.g. what's in it for me?) is key towards gaining their buy-in and acceptance. For a variety of reasons, each stakeholder has their own desires, wants and needs that they operate with. It is important to understand what these are and how they may relate to the ITSM Vision and Transformation Program overall.

Addressing these accurately with each Stakeholder will result quite favorably when it comes to overall acceptance of the ITSM solutions. The more that everyone can see a stake in the solution, the more successful you will be at leading your organization towards it. Likewise, those that do not see a stake in the vision will soon move onto other priorities and may even work against you if they feel the solution runs counter to their own objectives and needs.

Determining the needs for each stakeholder will be done mostly by meeting with them individually and flushing out their concerns and interests.

The table below provides some guidance towards this. It lists example types of stakeholders and identifies possible wants and needs that they typically have.

Type	Typical Wants and Needs
Customers	• Value for money spent if charged for IT services • Better service – can do work faster and more efficiently • More attention from IT • More influence over IT plans and activities
Senior Management	• Meeting corporate quality and improvement initiatives • Improving service • Getting ready for a major business initiative or change • Preventing calls from other business units about IT service issues • Demonstrating compliance with regulatory requirements • Understanding of costs for providing IT services
Middle Management	• Meeting business unit targets and objectives • Keeping staff and direct reports motivated • Preparing for a major business initiative or change • Preventing calls from upper management about IT service issues • Keeping IT costs in line with budget targets and objectives
IT Delivery and Support Staff	• Recognition for work performed • Eliminate firefighting • Working smarter instead of harder and longer • Reducing time spent on non-value work
IT Technical Architecture Staff	• Major support for IT Architecture standards and initiatives • Major support for IT governance • Recognition for technical solutions that support desired processes and business objectives • Increased visibility for architecture role and activities
Sponsors	• Return on investments made in ITSM • Successful implementation program • Meeting program objectives
Users	• Get work done without service disruptions • Get work done without spending time dealing with IT technical issues and problems • Better communications about service outages and root causes for them

Type	Typical Wants and Needs
Vendor Suppliers	• Better relationship with your business organization • Opportunity to increase services or length of service contracts • Clear definition of service roles and responsibilities • Clear roles and responsibilities for handling problems and incidents to reduce finger pointing and blame
IT Development Staff	• More efficient process for placing IT solutions into production • Increased business satisfaction with new IT solutions • Clear roles and responsibilities for handling problems and incidents to reduce finger pointing and blame between development and operations • Ensure deployment and operational considerations baked in as solutions are developed versus scrambling at the end before deployment
Change Agents	• Recognition • Chance to learn something new • Work on interesting projects • Make a positive difference in the organization
Stakeholder Initiatives	• Gain increased support and sponsorship for initiatives already underway • Share like ideas and improve solutions already under design • Increased influence and visibility for the initiative being worked on
Champions	• See desires turn into actions that make a positive difference • Gain visibility and recognition for opinions and foresight • Desires to see things done in the best possible way

In your Stakeholder Map, create an additional column for the stakeholder needs. For each stakeholder, inventory their needs.

Name	T.	Typ.	Pri.	Acc.	Inf.	Needs
Jim Jones	...	....	....	....	....	• Improving service • Getting ready for the new merger
John Smith	...	....	....	....	....	• Increased business satisfaction IT
Becky Thomas	...	....	....	....	....	• Get work done without service disruptions • Reduce time spent with IT technical issues • Better communications about service outages

An important source not to be overlooked is the Customer Needs Survey, ITSM Focus Survey and the ITSM Benefits Survey which are available to you in ITSMLib. A review of the survey results can help you determine where people put a priority on what they would like to see and get out of the ITSM Program.

Defining a Communications Plan

The Communication Plan is the key document that will be used to direct and guide the ITSM Communications program. It will be constructed, managed and executed by the Organization Change Team. The plan itself should operate as an independent track of effort done in parallel with implementation of ITSM tools and processes.

The plan itself defines the specific communications to be prepared and delivered, the timing for each communication, and the overall series of communications campaigns. In this book, we will consider communications to be delivered as a series of communication events. Events can be items such as:

- Training sessions

- Newsletter publications

- Presentations at key staff meetings

- T-Shirts or other bling promoting the ITSM program

- Launch of an ITSM Program website

- Agenda item at a leadership meeting

- Posters planted on office walls or cubicles

- ITSM user awards

- Anything else where ITSM related messages are being delivered

Within the plan, events are grouped into campaigns. A campaign is a series of related messages or milestones. More on this later. For each campaign, the plan identifies the events to take place. For each event it identifies the communication media, timing, applicable messages for that media, and audiences to be targeted.

The details for building communication events and campaigns will be covered in later chapters. For now, it is important to be aware of what a communication plan is and its importance. An example outline for a plan is shown below:

ITSM Communications Plan (Example)

1.0   Overview
2.0   ITSM Program Vision
3.0   Target Audiences
4.0   Stakeholder Analyses
5.0   Campaign Definitions
6.0   Event Definitions
7.0   Key Messages
8.0   Delivery Channels
9.0   Campaign and Event Schedules
10.0   Campaign Work Plans
11.0   Communication Cost Model

A brief summary of each section in the plan is as follows:

Overview

Presents why the plan has been produced and what readers can find in the document.

ITSM Program Vision

Summarizes the vision of the ITSM program, business benefits, target timeframes and the purpose of the program.

Target Audiences

Identifies the types of stakeholders that are involved or will be impacted by ITSM campaigns

Stakeholder Analyses

Summarizes the stakeholders involved and their information needs. The Stakeholder Map can be included with this to provide detail.

Campaign Definitions

Lists each campaign, the key messages, success factors, planned outcomes and timeframes. Also identifies which events will be conducted as a part of each campaign.

Event Definitions

Lists the communication events, their planned delivery timeframes, planned audiences, delivery media to be used, which campaigns they belong to and which messages they support.

Key Messages

Summarizes the key messages to be delivered by stakeholder. Indicates what events and campaigns messages will be delivered in.

Delivery Channels

Describes how messages will be delivered. Maps the channels to the events that will use them. While some channels are obvious (e.g. Email), others may need to be described (e.g. Mystery Box sent to each IT Support Person).

Campaign and Event Schedules

Lays out the events and campaigns over a calendar timeframe.

Campaign Work Plans

Presents work plans showing how events and campaigns will be built, monitored and managed as they are conducted. Shows build and execution tasks, who is responsible, estimated level of effort and task timeframes.

Communication Cost Model

Presents a model of costs for building and running the communications part of the ITSM program.

Chapter 9

Stage 2 – Strategize Campaigns

In this stage, the key objectives will be to identify the key communication events and campaigns that will take place. This will be done by first identifying what information needs exist for each type of stakeholder identified. Information needs are then grouped (usually by 2-3 needs per group). The groupings become the basis for campaign events which make up the campaigns which are then placed into an overall timeline and schedule.

Approach Overview

For this stage, the following activities should take place:

Step	Action
01	Identify what needs to be communicated for each stakeholder type
02	Identify the campaigns
03	Map information needs into the campaigns
04	Map each stakeholder type into each campaign
05	Identify timelines and schedules for campaigns
06	Identify campaign measurements and success factors
07	Clarify the vision
08	Establish operating principles for communications

Key outputs from this effort include:

- Inventory of communication needs

- Inventory of communication campaigns

- Campaign estimated timelines and schedules

- Campaign success factors and measurements

- Clarified vision

- Communications operating principles

Identifying What Needs To Be Communicated

In this step, go down your list of identified stakeholder types. What kinds of things do they need to know to be successful? What kinds of questions will they have that need to be answered? Construct a list for each type of stakeholder you have and update your Stakeholder Map with this information.

The table below presents an example of information needs that was gathered for an actual IT organization:

Stakeholder	Program Communication Needs
Senior Management	Overview of ITSM Program Program roadmap, goals and timing Program rationale and key benefits Relationship to other IT initiatives Program leadership and staffing IT Business and IT Support services Service portal Program status and accomplishments Program costs
IT Middle Management	Overview of ITSM Program Program goals and timing Program rationale and key benefits Relationship to other ISD/ITS initiatives Program leadership and staffing Program status and accomplishments Departmental responsibilities Impacts on their department and timing Impacts on their personnel and timing Who to contact for answers Ways they will be informed and involved Status of ITSM training/skills attainment Knowledge bases available to them KPIs, CSFs and Management Dashboards What's in it for them Service portal ITS Business and IT Support services Operating roles Accessing program information resources SMS Tooling Change Business Impact SMS Tooling Training

Stakeholder	Program Communication Needs
IT Support Staff	Overview of ITSM Program Program goals and timing Program rationale and key benefits Relationship to other IT initiatives Program leadership and staffing Program status and accomplishments Their personal responsibilities Impacts on their department and timing Impacts to their position and timing Who to contact for answers Ways they will be informed and involved What's in it for them Service portal IT Business and IT Support services Knowledge bases available to them Accessing program information resources Tool Training
Business Liaisons	Overview of ITSM Program Program goals and timing Program rationale and key benefits Relationship to other IT initiatives Program leadership and staffing Program status and accomplishments Their personal responsibilities Impacts on their department and timing Impacts to their position and timing Impacts to the departments they represent Who to contact for answers Role of IT business liaison Service portal IT Business and IT Support services Ways they will be informed and involved What's in it for them How to access information resources Business Impact on their Customers
End Users	Overview of ITSM Program IT Business and IT Support services Service portal How to do business with IT Program roadmap, goals and timing Program rationale and key benefits Program accomplishments What's in it for them Role of IT business liaison Who to contact for answers Ways they will be informed and involved Business Impact on their Department

Stakeholder	Program Communication Needs
Suppliers	Overview of ITSM Program Program goals and timing Program rationale and key benefits Program status and accomplishments Impacts on the services they deliver Changes in expectations, delivery handoffs Impact on existing contracts Their responsibilities Who to contact for answers Ways they will be informed and involved What's in it for them Business Impact on their Services
Service Owners	Overview of ITSM Program Program goals and timing IT Business and IT Support services Service portal Program rationale and key benefits Relationship to other IT initiatives Program leadership and staffing Program status and accomplishments The role of a service owner Coordinating with other service owners Coordinating across other departments Their personal responsibilities Who to contact for answers Ways they will be informed and involved What's in it for them How to access information resources Tool Training

Identifying the Campaigns

Campaigns will ultimately consist of large chunks of communication events that are given over time. Campaigns can be considered milestones in the communication effort that each achieve a level of change progressing towards the new solutions being put into place.

In designing campaigns, think in terms of a story that is taking place. There will be a beginning, middle and end set of campaigns for each type of stakeholder. A generalized "story" might look like this:

1. Stakeholders first learn that something is about to change

2. Stakeholders then learn new ways of working

3. Stakeholders prove they understand how to work in those new ways

4. Stakeholders then adopt and transition to those new ways of working

5. Stakeholders receive help and support until they feel comfortable with those new ways of working

Each step in the above "story" can be translated into a communications campaign. An example using the above might look as follows:

Campaign #1 - Awareness

Stakeholders must be made aware that changes are about to take place. This campaign will include events such as awareness sessions, company announcements and a newsletter publication.

Campaign #2 - Learning

Stakeholders must know how to use new processes and tools that will support the changes taking place. This campaign will include events such as training sessions for the new processes and tools.

Campaign #3 - Adopting

Stakeholders must demonstrate that they can use new processes and tools with little help or support. This campaign will include events such as tests and exercises that stakeholders can do on their own.

Campaign #4 - Deploying

Stakeholders are participating and aware that ITSM solutions are now being deployed and use of these will start within the next business day. This campaign might include events like a deployment walk through, deployment checklist gate, or deployment rehearsal.

Campaign #5 – Early Life Support

Stakeholders are provided with extended help and support post implementation to minimize confusion and address any issues they may have. This campaign will include events such as optional training sessions, "ask the expert" sessions, feedback sessions, and a hotline for quick answers to stakeholder questions.

Note that with each campaign shown above, a tag name (e.g. Awareness, Learning, Adopting, etc.) has been given to provide a more meaningful name to the campaign.

Mapping Communications to Campaigns

At this point you have a list of campaigns and a list of information needs. Sort the information needs so you can remove any duplicates. Now go ahead and map each information need into one of the campaigns you have identified. If you find there is an information need that does not fit into any of the campaigns, consider:

- Adding a new campaign for that information item

- Changing the scope of a campaign you already identified to include the new item

The table below presents an example for this:

Information Item	Campaign
Overview of ITSM Program Program goals and timing IT Business and IT Support services Service portal Program rationale and key benefits Relationship to other IT initiatives Program leadership and staffing Ways they will be informed and involved What's in it for them	#1 Awareness
The role of a service owner Coordinating with other service owners Coordinating across other departments Their personal responsibilities Who to contact for answers Tool Training	#2 Learning
Program status and accomplishments How to access information resources	#3 Adoption

Identifying Campaign Timelines and Schedules

Once the campaigns have been identified, estimate and determine the timing for each one. When will each one start? When will each one end? Place these in a timeline with dates. This might look something like the following:

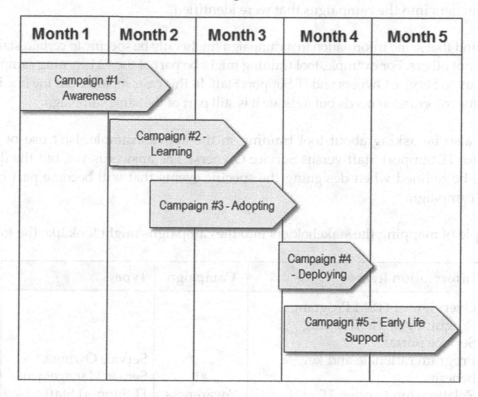

When estimating the timelines, consider:

- Number of stakeholders in each campaign that need to be communicated to

- Timing with implementation of ITSM tools and processes

- Other business events that might impact the schedule such as periods of time where IT is not allowed to make changes or a key change in senior leadership

- Employee vacations, key meetings or other events where stakeholders may not be available to participate in the campaigns

- ITSM project constraints such as deadlines to avoid further licensing costs of old tools, consolidation of an IT department or rollout of a major new application solution

At this point, the schedule is high level and draft. The schedule may change as you elaborate definition and design of the events that will take place within each campaign.

Mapping Stakeholders into Campaigns

So we now have our information needs and what campaign the information will be communicated in. The next step is to identify which stakeholders will participate in which campaigns. Looking back at the information needs identified for each stakeholder type, map the stakeholders into the campaigns that were identified.

You may find that some information in a campaign might only be specific to certain stakeholder types and not others. For example, tool training might be part of the #2 Learning campaign but only relevant to Service Owners and IT Support Staff. In this case, separate out the line items for these unique information needs but indicate it is still part of the same campaign.

You may also be asking about tool training in the above example. Isn't use of the tool different for IT Support Staff versus Service Owners? The answer is yes, but the details of these will be defined when designing the specific events that will become part of the #2 Learning campaign.

An example of mapping the stakeholders into the campaigns might look like the following:

Information Item	Campaign	Type
Overview of ITSM Program Program goals and timing Service portal Program rationale and key benefits Relationship to other IT initiatives Program leadership and staffing Ways they will be informed What's in it for them	#1 Awareness	Service Owners, Senior Management, IT Support Staff, Business Liaisons
The role of a service owner IT Business and IT Support services Coordinating with other service owners Coordinating across other departments Their personal responsibilities Who to contact for answers	#2 Learning	Service Owners
Tool Training	#2 Learning	Service Owners, IT Support Staff

Information Item	Campaign	Type
Program status and accomplishments How to access information resources	#3 Adoption	Service Owners, Senior Management, IT Support Staff, Business Liaisons

Measuring Campaign Success

Here you will indicate the success factors for each campaign. These should related to the outcomes you want for each campaign and must be measurable. Campaigns will be considered a success if all their success factors have been achieved.

The following presents an example of how you might measure the success of the example campaigns that have been presented.

Campaign #1 - Awareness

Key targets for this campaign are:

- 100% of directly impacted personnel will be aware of the ITSM Implementation Program.

- Over 90% of managers will be able to explain the ITSM Service Vision and ITSM Program objectives in their own words.

- Over 80% of impacted stakeholders will be able to explain the ITSM Implementation Program and its key objectives.

- 75% of impacted stakeholders will have voluntarily participated in at least one Brown Bag Lunch presentation about the ITSM Implementation Program.

- Over 80% of Managers and Executives rate ITSM (via survey) as necessary to the long-term positioning and survival of the IT organization

Campaign #2 - Learning

Key targets for this campaign are:

- 100% of directly impacted stakeholders know the required ITSM training schedule applicable to their position

- Over 95% of stakeholders have attended their required training events

Campaign #3 - Adopting

Key targets for this campaign are:

- Over 95% of stakeholders have completed their online ITSM tests and exercises

- Less than 10% of all impacted stakeholders have registered complaints or issues with the ITSM Program

- 90% of all inbound communications are answered (or acknowledged, if the answer must be researched extensively) within 1 business day

Campaign #4 - Deploying

Key targets for this campaign are:

- 90% of all inbound communications are answered (or acknowledged, if the answer must be researched extensively) within 1 business day

Campaign #5 - Supporting

Key targets for this campaign are:

- 90% of all inbound communications are answered (or acknowledged, if the answer must be researched extensively) within 1 business day

- 90% of requests related to new solution issues and deficiencies are satisfied within 3-5 business days

- Over 70% of impacted stakeholders can describe at least one benefit or success story associated with the ITSM Program

Clarifying the Vision

For this step, ensure that there is a clear unifying set of vision statements and messages. Prior to this, senior leadership may have already established a vision. If so, then review it for clarity and purpose. Make sure it is compelling enough to build campaigns over. If a vision is unclear or not established, then use this part of your effort to put one in place. The program will fail without it.

IT organizations easily get caught up in delivering services. They will cast ITSM aside under pressure to make sure those services are safely delivered. Only under pressure (e.g. major outage, loss of business, bad publicity, management dictate) will they look for change or view things differently.

IT operations, especially likes things cookie cutter and not to change for good reasons. They are change averse. They recognize that when the delivery machine is disrupted, bad things can happen. Their avoidance of change comes from natural need to formalize IT operations so they will reliably and systemically support growing customer demand.

In addition, you can't take an IT organization down to retool. New ways of working and operating must be put into place, all while maintaining organizational and operating continuity.

For these reasons, 3 things must be in play or people will not follow:

1. People must feel that not everything is working as it should. There is discontent with how certain things are working. Maybe concerns about future events or conditions that will threaten what is being done today. If they feel things are going okay, they will not change.

2. People see viable alternative to ways of operating that address the concerns or feelings they have about the current state. They must believe that these alternatives offer something of value to them personally as well as the IT or business organization at large.

3. People must be able to see and agree a path or set of steps that progress forward towards those alternative ways of operating. Otherwise they will resist and not change ("Great idea…won't work here…").

The vision must be crystal clear on all 3 points. If any of them are not in place as part of your vision, there will be great risk of resistance and eventual program failure.

So what's provided by a vision?

Hard elements:

*   The issues being addressed

*   The changes being put into place to address those issues

*   The means by which the organization will progress towards those changes

Soft elements:

*   Clear understandable focus and conflict free mission for each level of the organization (e.g. senior leadership may see different objectives that line staff, but the objective of one does not harm the other – in other words, benefits for all)

*   Matches the values of individuals as well as the IT and business organization

*   Visible commitment by senior leadership and management

*   Respectful of those delivering the services

As you craft your vision, make sure every element, hard and soft, is being addressed with it.

The table below presents some examples of how the focus and mission can differ at various levels of the organization:

Level	Examples of Focus
Senior Leadership	• Cost efficiency • Lowered operating risks • Security • New opportunities • Growth opportunities • Business unit satisfaction and respect
Middle Management	• Operating efficiency • Meet operating targets • Span of control • Meets budget constraints • Makes their unit numbers
Support Staff	• Job recognition • Avoid overtime work • Pay and bonus • Respect for accomplishments
Suppliers	• Close a sale • Make a bonus or commission • Don't waste time on accounts that will not yield sales • Deliver good service for accounts that have been sold • Meet service targets • Avoid service penalties and bad press
Business Users	• Meet their business outcomes • Enhance their business functions • Open new opportunities • Get into new markets • Increase sales

When crafting a vision, the wrong questions are asked too often:

- How big should the organization be?

- How do we get mean and lean?

- Should we centralize, decentralize or re-engineer?

- Should we get better trained or certified?

These are operational questions that are focusing on solutions without a clear picture of the where the organization should head. They are cherry picked based on intuition, guesses and other means but not tied to any specific objectives.

More powerful questions to consider when crafting a vision might be:

- How do we make our workforce more innovative and productive?

- How can our organization be more responsive to unexpected shifts in the marketplace?

- How should we manage and support rapid changes in new technologies?

- What is the best process for redesigning and revitalizing our IT organization?

- How can we better serve our customers – internal as well as external?

- How do we increase synergy across multiple teams and reduce fragmentation?

- Given the direction of IT in the industry, how do we best adapt and change to meet that direction?

The Vision statement itself is extremely important. This will be the flag that everyone will rally around that serves to stimulate excitement and interest in what is to be achieved. It does not need to be too detailed. If those reading or hearing it become excited and ask "tell me more…" it will have served its purpose.

Several rules on Vision Statements need to be considered:

- They should be crystal clear and one sentence if at all possible.

- They should identify a stated goal in generic terms

- They should not be cluttered with statements about how the Vision will be achieved

- They should not include specific product names or standards

Here are some good examples of ITSM Vision statements that have been used at other companies:

- To be the preferred (your company name) IT service partner.

- We will set the industry standard for how to operate a global IT infrastructure that is admired around the world – most importantly - by our customers as a reason to buy products and services from our company.

- We support our business goals of transformation into a leading company by providing the highest quality IT service infrastructure that delivers the most reliable, accurate and secure information services in the world.

- We will always protect our corporate investments in IT solutions and infrastructure by ensuring that they will be deployable and operable on a day-to-day basis at acceptable cost.

- We will deliver a range of responsive and integrated services that will meet, and where appropriate, exceed our customers' expectations.

- We will set the standard for how to operate a set of infrastructure management services that is admired around the world – most importantly - by our associates who will proudly use our services

- We will create and operate a set of infrastructure management services parallel to no one else in our industry

Here are some not so good examples of ITSM Vision statements that have been used at other companies:

- We will use ITSM to drive the standard for how our IT services will be delivered.

- We will achieve an IT Service Management CMM level of 4.0 or higher.

- Through improved ITSM alignment and process maturity, we will be able to:

  – Support and further corporate mandates and objectives

  – Effectively utilize resources through the clear definition of roles and responsibilities

  – Enable us to absorb growth with a virtually flat staffing level by way of achieving process excellence

  – Align service delivery with service level requirements

  – Implement changes faster, improving business agility

  – Reduce the volume of Incidents resulting from Change

  – Improve capture and utilization of operational data for proactive purposes

  – Provide efficient and effective response to service outages

  – Reduce the number of service outages

  – Resolve issues faster, minimizing business impact

  – Analyze and intercept Incidents and Problems before they impact our customers

In the above examples, the first two really address how a vision will be achieved, but does not state what the vision is. The last example, which is fairly typical, states some good things, but is far too complicated and long to describe to someone in a 30 second elevator speech. Just reading it may cause some eyes to glaze over.

Communication Strategies That Seem To Work

In formulating your campaign approach, consider some of these high level strategies taken from many other ITSM transformation efforts:

- You can almost never over communicate. Most people need to hear a message 5-7 times over and over before they really understand it.

- Almost no one fights against a solution they feel they have had a part in developing. Seek the broadest participation from all stakeholders who will have to build or use the ITSM solutions being developed.

- Keep an open atmosphere for all ideas, concerns and complaints. It helps people to air these out. Many times they come from confusion or misunderstanding as to what the program is supposed to accomplish or how it will work. All ideas and arguments should be welcome up to the point of deployment.

- Create time needed to effect change without forcing IT to fit an unrealistic timeframe. An industry rule is that most people will reject changing if the effort to do so exceeds 10-15% of their working day. Forcing change faster than it can be absorbed can not only increase resistance, but also put the program objectives at risk.

- Recognize that there are some periods of time where the organization needs a breather, especially after a major transformation effort has finished. It's possible that people want some time to absorb and get used to the new systems before starting in on improving them further.

- Make sure you're communications are based on authoritative sources, numbers, data and facts. Keep reactions, feelings, and politics out of discussions and decisions.

- Get the right scope of change. Only changing a small slice of the organization can result in alienation where the rest of the organization feels disconnected from the group doing the change. "It's just for that group. They're not really part of us" they will say.

- Change needs to encompass the entire organization. All staff, functions, management levels and diverse geographic locations that will be impacted by the change.

- Commit the best people possible. Loading the communications team with dead wood or people that don't fit into other parts of the organization not only puts your program in jeopardy, it sends a message that this effort is not really important.

- Utilize a "time boxing" approach if the IT staff is too busy to focus. With this, identify a set time each week to solely focus on ITSM activities (e.g. "...every Friday from 1-2pm each week we will meet in the war room to solely focus on our ITSM Program..."

- Be aware that just because people nod their head or don't dissent in meetings, doesn't mean that they really accept the ITSM direction or solutions being built.

- Avoid the "dictator" approach as much as possible. If people feel this is "being done to them", they will resist over time and the whole program will fail.

- Adoption and acceptance are like waves in the ocean. Once people see the wave coming, they will get on board. Utilize change leaders (who can be at any level in the organization) that can get the waves started and soon the rest of the organization will follow. Also note that the opposite can happen which can eventually kill your program (e.g. "…this will never work here…" or "…management is not really behind this…").

- IT senior leadership needs to be seen and heard. This can be as simple as a 2 minute appearance at a key communication event explaining why this is important. Support staff will quickly pick up on negative outcomes if they never see the leadership involved in this.

- Recognize the culture for what it is. Everyone agrees that best practices for IT are good, but they won't change or move on anything if the culture doesn't allow it. This is why the Awareness step is so important.

- Recognize the management mission for what it is. You may be focusing on all the good things ITSM can bring to the organization in terms of quality, less downtime, and greater customer satisfaction. Senior leadership may only want to consolidate the Service Desk and reduce headcount. Make sure you truly understand the management mission.

- Never assume that a stakeholder truly understands or accepts what you are communicating. Try to step in the shoes of the person and see how they are looking at the world. Not everyone's view of the solution will align with ours.

Campaign Communication Principles

As part of the strategy, consider setting some overall communications operating rules. Below are some examples:

- To ensure IT managers and leadership feel that they are well informed, communications to them will precede communications to the people who work for them.

- Communication delivery will be done by the most credible source(s) for that particular message, including executives and other members of the management team.

- Each campaign will be based upon specific objectives. At the end of each campaign, these objectives will be evaluated, and follow-up measures will be taken as appropriate.

- The Communication Program will focus on delivering critical messages up to nine times to large groups of affected people. No message will be delivered to any audience fewer than three times.

- Campaign communications will be written to honestly portray important information, even if it will be viewed negatively.

- No communications will isolate, insult or humiliate any person or group, even if that person or group has had a past history of poor performance.

Chapter 10

Stage 3 – Design Campaigns

In this stage, the key objectives will be to identify, flesh out, and design the campaign events in detail. This includes design of any materials to be used and how the campaigns will be delivered (e.g. training class, email blast, company meeting, etc.). Stakeholders are then assigned to each campaign and a means for tracking and monitoring attendance and progress of the campaigns is identified.

Approach Overview

For this stage, the following activities should take place:

Step	Action
01	Identify campaign communication channels
02	Identify campaign events
03	Link delivery channels to campaign events
04	Identify who will conduct events
05	Design campaign event details
06	Assign stakeholders to campaign events
07	Design campaign management mechanisms
08	Update communications plans and schedules

Key outputs from this effort include a design package with:

– Campaign events

– Communication channels that will be used with each event

– Stakeholders assigned to events

– Identified resources to build and conduct events

– Designs for event feedback and progress reporting

– Updated communications plans and schedules

Identifying Communication Channels

Communication channels are the vehicle by which communication is transferred from those making changes to their stakeholder base. These vehicles can take a wide variety of forms. Some examples include:

- Meetings

- Blogs

- Workshops

- Presentation deck with common talking points

- E-Mail Blasts

- Voicemail Messages

- Lunch and Learn

- Train-The-Trainer Support Staff

- Newsletter

- Management Memo to Staff

- Presentations and staff meetings

- Public or internal websites

- "Ask Your Question" voicemail box or website

- Branded Item (e.g. Cups or T-Shirts)

- Signs and Displays in the workplace

- One-on-one conversations

- "Day in the Life" video

- Education and training sessions

- Rewards and recognition

- Surveys

- Social media

- How-To Videos

- Contests and games

- Off-site forum or symposium

- Guest speakers

Also consider geography with this. Are stakeholders all in one place? Spread out across the country? Do they have time to personally attend communication events or would they prefer to do these on their own time? You may choose different vehicles depending on your organization and the strategies you choose.

In short, be creative! The more engaging the vehicle is, the better the communication will be received. You should also align the vehicle with the tone of the communication that is needed. For example, holding contests and games might not work for an executive meeting where there are performance concerns about service delivery.

Identifying Campaign Events

Campaigns consist of one or more events that take place all in the service of the campaign goals and messages. In this step, you will identify what events you want to take place. A key input for this is to look at the stakeholders linked to each campaign that was done earlier. High level considerations might be as follows:

Can one event serve all stakeholders in the campaign or will different kinds of events be needed for different kinds of stakeholders? Using the Awareness Campaign example, can one event be the same for executives as well as line staff? Or should separate events take place for each of those audiences?

Will there be just one event to get the messages across or should multiple events take place over time? Will there just be one announcement to make everyone aware of the ITSM Program and then everyone moves on versus a series of events over time to get everyone more used to the idea?

Whether all messages are to be communicated in a single event or through multiple events. For example, when communicating the new support processes do you cover everything there is to know incident, problem and change in a single event or split into separate events for each process?

Whether the entire stakeholder audience will attend a single event versus having repetitions of events to handle larger audiences. For example, you may work in an IT organization with hundreds of support staff. Do they all get training at once? Will multiple sessions be offered with only 20-30 attendees (in this case necessitating 10-20 sessions)?

Whether all details are to be communicated in a single event versus multiple events. For example, holding 2 events to cover the Problem Management process – one for how the process works and one for how to use the new tool to support the process.

Take a look at each campaign identified earlier? What kinds of events should take place for each campaign? Use the stakeholder map and campaign information developed earlier. You might lay this out like the following example:

Campaign	Event	# Times	Stakeholder	Messages
#1 Awareness	# 1 Company Blast	1	All	• Program Goals • Program Overview
#1 Awareness	#2 Program Detail	8	IT Support Staff	• Program Goals • Program Overview • What's in it for them • How the Program will operate
#1 Awareness	#3 Program Overview	2	IT Managers	• Program Goals • Program Overview • What's in it for them • How the Program will operate • Monitoring Program Success • Program Measurements
#2 Learning	#4 Incident Management Process	6	IT Support Staff and IT Managers	• Logging incidents • Handling incident escalations • Resolving incidents
#2 Learning	#4 Incident Management Tooling Administration	2	Service Desk Staff	• Logging into the tool • Receiving an incident ticket • How to escalate tickets • How to close tickets

With this, we have taken an important step towards linking the campaigns with their events, who will be communicated to by each event and what key messages will be communicated in each event.

Linking Delivery Channels to Events

For this step, identify the communication vehicles that will be used for each event. Is the event a meeting? A workshop? A newsletter announcement? The table below provides an example for how you might formally identify which vehicles will go with which event:

Delivery Channel	Event #1	Event #2	Event #3
E-Mail Blast	X		
Voicemail Blast	X		
Newsletter	X		
Weekly Staff Meeting		X	
Cups and T-Shirts		X	
Workplace Signs	X		
Training Classes			X

Identifying Who Conducts Events

In this step, identify the strategy for who will conduct each event. Example strategies might include:

- In-house staff that will lead events

- Outside consultants

- In house or external training providers

- ITSM Communications Team members

- Middle Managers

- Executive leadership

Designing Campaign Event Details

In this step, details for each campaign event are designed. This is where the hard work comes in. Examples of items that need to be designed include:

- Materials that will be used with each event such as training guides, test exercises, and presentation decks

- Whether 3rd party resources will be needed to build and/or conduct events (e.g. will a training provider be used with pre-built training content?)

- Scripting items such as what a management announcement might say or what will be shown and said in a video

- Logos for the program

- Communication templates to give the program a common look and feel (e.g. presentation deck or email templates)

- Training and workshop facilities (e.g. classroom physical makeup, audio-visual aids, etc.)

- Program web sites

- Program support mechanisms such as hot lines or responses to emailed questions

- How attendance at events will be monitored, tracked, and reported on

- How feedback from events will be captured and processed

- How stakeholders will register and commit to attending events

Assigning Stakeholders to Campaign Events

Up to this point, we have identified which type of stakeholder will participate in which events. For this step, we need to now identify, by name, who will attend each event. Important considerations include:

- Making sure you have a complete inventory of names tied to each stakeholder type and not leaving anyone out

- Making sure the audience size for each stakeholder type matches the estimates made earlier

- Checking to make sure people are not overly assigned to events. For example, if John Smith is booked to attend 30 training events, there may be an issue. If this happens, you may need to revisit your event strategies.

- Estimate of time impact that will each person will incur. For example, an E-mail blast might be seconds or minutes, but a training session might be 2-3 hours.

- Any well-known future changes that impact whether a person should really participate. Examples might be a senior executive who is leaving or retiring – should this person spend time on this? Another example might be a support team that may be eliminated, outsourced, or moved to a different organization.

- Geography may also be a consideration. For example, are all stakeholders at one location or are they spread out across the country?

Designing Campaign Management Mechanisms

With this step, design the details for how campaigns will be monitored and managed. Key areas that need to be designed here include:

- Feedback mechanisms such as survey questions.

- How feedback will be captured, scored, tabulated and summarized. This could be done on spreadsheets, but there may be a desire to use more sophisticated tools such as an online survey tool.

- Records and tracking for events such as training.

- Registration systems that stakeholders can use to confirm their attendance at events such as an online sign-up sheet.

- Reports and scorecards used to show progress and success attainment with communications activities.

- Repositories to hold communications artifacts such as presentation decks, copies of announcements, training materials, or status reports.

- Standards such as naming of artifacts (e.g. "Is this deck the overview presentation given to executives on March 3rd?"), templates or common look and feel.

Interesting Ways to Communicate

There is no end to the creativity that you can employ to communicate with your stakeholders and help them adopt and use the new solutions you are building. Listed below is an inventory of some interesting approaches that other IT organizations have been taking to help their organizations transform:

Event	Description
Open Demo	Establish a war room set up with access to the new tools and processes that people can go into at any time to try out the new solutions, ask questions, or raise concerns.
"Doctor" ITSM	Appoint someone to act as the role of a "doctor" complete with surgical outfit or other dress to hold an open meeting for support staff to discuss their specific issues and concerns. Make sure food is served at this event. Any question asked is fair game.
ITSM Scavenger Hunt	Place hidden information (e.g. process item, incident resolution, knowledge item) within your new ITSM tools and offer a prize to the first group of individuals that can find them.
Find-The-Root-Cause Contest	Present a series of mock incident tickets and see if anyone can identify the underlying root cause.
ITSM Crossword Puzzle	Crossword puzzle with recognition and a prize for the first group of people that can solve it.
Mock CAB Meeting	Conduct a mock CAB session with members to walk through the CAB meeting activities. This can also be done ahead of time with the old tool RFCs but handled with the new process in the meeting.

Chapter
11

Stage 4 – Conduct Campaigns

In this stage, the key objectives will be to build and execute on the campaigns and events designed in the earlier stages. This includes key activities like acquiring delivery resources that will support and conduct the campaigns, building campaign materials and artifact, and scheduling the campaigns.

Approach Overview

For this stage, the following activities should take place:

Step	Action
01	Procure and assign campaign delivery resources
02	Build monitoring and feedback mechanisms
03	Develop and build campaign event materials
04	Publish campaign materials
05	Finalize campaign schedules and timelines
06	Update communications plans
07	Formally announce start of campaigns
08	Conduct campaigns

Key outputs from this effort include:

- Campaign materials and artifacts such as presentation decks and training guides
- Installed campaign feedback and monitoring infrastructure
- Updated communication plans
- Finalized campaign event schedules and logistics

- Communications to stakeholders about campaign events and when these will take place

- Conducted campaign events (e.g. workshops, training, announcements, etc.)

- Event feedback and progress reports

Establishing Campaign Delivery Resources

In this step, finalize the resources you need to build and carry out the campaigns. Key activities for this can include:

- Procurement and contracting activities to obtain 3rd party resources needed to build, train and conduct campaign events

- Reassigning internal work staff to build, train and conduct campaign events

- Finalizing the roles and responsibilities of the Organizational Change team during this stage of the effort

- Finalizing the roles and responsibilities of IT support staff and other stakeholders that will actively participate in building and running campaign events

Publishing Campaign Event Materials

At this point, designs for materials and artifacts to be used in campaign events has been designed. Activities at this stage are to build and publish those artifacts. Examples of items that may need to be built can include:

- Training guides and classroom materials

- Email campaigns building out which emails will be sent when and with what information over a progressive period of time

- Writing content for newsletters or other publications

- Publication of how-to guides

- Development of how-to videos or other media

- Development of talking point presentations

- Subscription lists for email blasts and announcements

- Web sites for self-help, training or other information

- Facilities to register for training or other events

- Reservation of physical building space for training or other events

- Arrangements for technical support for online sites, technical class training or audio-visual equipment

- Arrangements for conferencing facilities and online conferencing meetings

- Scripts and talking points for senior leadership

- Event tracking and reporting infrastructure

- Build out for special events that may be planned such as a company party or an open "try it out yourself" event

- Hardware, software and physical facilities to hold special demo events

- Lab work exercises to that stakeholders will do on their own to get independent hands-on experience with new tools and processes

- Laptop or device instance for needs where events may need to be conducted outside the traditional workspace environment. The device would have to be populated with the new tool instance, network connection and any presentation decks or supporting materials

Scheduling Campaign Events

Once materials are close to ready, finalize the scheduling of campaign events. The campaign schedule will become a key part of your communication plan. Below is an example of one actually used to deploy a new Change Management process and tooling solution:

Event ID	Communication Event	# Involved	Delivery Approach	Timing	Duration	Comments
01	Process Announcement	200-230	Email Blast	2/18	n/a	Director sent, Program Announcement
02	What's Up Notification	200-230	Email Blast	2/19	n/a	Change Manager sent, what's coming up
03	Time-To-Schedule Notification	200-230	Email Blast	2/20	n/a	Change Manager sent, request to schedule training for awareness
04	Process Awareness	200-230	Recorded Webinar and Classroom	2/23 – 3/20	1 Hour	Series of pre-scheduled recorded webinars and classes – attendees sign up which one they will attend
05	Specific Role Duties – Service Owners	20-25	Classroom	2/23 – 2/27	½ Hour	Series of pre-scheduled classes – attendees sign up which one they will attend
06	Specific Role Duties – CAB Members	5-7	Classroom	2/23 – 3/6	1 Hour	Training Meeting
07	Specific Role Duties – ECAB Members	2-3	Classroom	2/23 – 2/27	½ Hour	Training Meeting
08	Time-To-Schedule Notification	200-230	Email Blast	2/26	n/a	Change Manager sent, request to schedule training for tool overview once they have finished awareness training
09	Tool Overview	200-230	Recorded Webinar and Classroom	3/2 – 3/31	1 Hour	Process review with tool demo, Series of pre-scheduled classes and a recorded webinar – attendees sign up which one they will attend

Event ID	Communication Event	# Involved	Delivery Approach	Timing	Duration	Comments
10	Progress Notification	200-230	Email Blast	Every Monday	n/a	Notification showing who is on track for training and who is not – goes to all stakeholders
11	Lab Work Notification	200-230	Email Blast	3/2 – 3/31	n/a	Notification and details for lab assignment - Change Manager sent, request to schedule lab work once they have finished tool overview training
12	Hands-On Tool Walkthrough	200-230	Lab Work With Optional Classroom	3/2 - 3/31	n/a	Stakeholders may complete lab work on their own time or optionally attend a scheduled class if they want some handholding (first 10 people to get lab work done get a prize?)
13	Tool Walkthrough – Service Owners	20-25	Lab Work With Optional Classroom	3/2 – 3/31	n/a	Stakeholders may complete lab work on their own time or optionally attend a scheduled class if they want some handholding
14	Tool Walkthrough – CAB Members	5-7	Lab Work With Optional Classroom	3/2 – 3/31	n/a	Stakeholders may complete lab work on their own time or optionally attend a scheduled class if they want some handholding
15	Tool Walkthrough – ECAB Members	2-3	Lab Work With Optional Meeting	3/2 – 3/31	n/a	Stakeholders may complete lab work on their own time or optionally attend a meeting if they want some handholding
16	Certification Note	200-230	Email Blast	3/2 – 3/31	n/a	Send stakeholders a certificate (or T-Shirt?) that they are now certified for Change Management, extra marks for service, CAB and ECAB members
17	Process Feedback	200-230	Direct contact, survey	2/23 – 4/10	n/a	Early life support for issues, survey monkey for feedback

Stage 5 – Enforce Change

In this stage, the key objectives will be to monitor and report on campaigns as they take place. Ongoing activities occur to manage adoption, deal with resistance and process feedback from stakeholders. The focus of this phase is to build sustainability of the change as the organization transitions and will sustain that change post transition.

Adoption and acceptance need to be continually monitored and managed even post transformation. There is a human tendency to easily slip back into old ways and habits because they are perceived as more comfortable. Keep a close watch! While things may seem to have transitioned successfully at the end of a transformation effort, pockets of small activities may revert back to the prior way of doing things. These pockets can grow slowly over time until your organization has slipped back before you know it.

Key outputs from this effort include:

- Ongoing communications to reinforce results and sustain the changes that have taken place

- Campaign results and measurements

- Feedback on ITSM program attainment

Approach Overview

For this stage, the following activities should take place:

Step	Action
02	Monitor and track campaign results
03	Provide early life support for campaigns
04	Monitor and manage adoption and resistance
05	Celebrate progress and successes
06	Obtain and act on feedback mechanisms

Measuring Campaigns

As you kick off your campaigns, take a measurement baseline. Re-measure this towards the end of the campaigns to check progress, adoption and acceptance. Be willing to change methods and/or what might take place in a campaign if it appears that the campaign is not working. A checklist of what to measure might include:

Attendance at campaign events.

Poor attendance may be indicator that people are not taking the change seriously or are not in agreement with the vision and stated solutions.

Training survey results.

Issuing a small survey at the end of key events or training sessions may be helpful in assessing both adoption and success of the events. Example questions might include: "How well do you feel you can apply what you have learned to your job?" or "Was this session helpful?"

Open Issues Log.

Maintaining an open issues log where anyone can freely add an entry. This should not be a complaints list, but more of an inventory of things that are not working or challenges that are not being overcome. Rate each entry as high, medium or low in terms of risk to the effort. Monitor and track the number of entries to determine if things are moving forward.

Random Skills Check.

Conduct random checks to determine if new skills are properly being transferred. The means for this can be through observations of people at work or audits of ITSM records such as incident, problem or change tickets. For example, a check that shows little or few problem tickets have been raised might indicate that people are not yet using the Problem Management process.

ITSM Meeting Event Attendance.

Conduct checks of attendance at ITSM meetings that are supposed to take place. Examples might include attendance at Change Advisory Board (CAB) meetings, major incident reviews, or service review meetings. Poor attendance or cancellation at these meetings could be a sign that adoption has not taken place.

Adherence to Campaign Event Schedules.

Here you can check the campaign progress itself. Have events taken place as planned? Are there delays that threaten to prolong the campaigns?

Operational Outcomes.

Checking operational metrics and measures post transition versus what they were pre-transition. For example, if incident counts were trending upwards before the campaigns started, are they now trending slower? Going down? Leveling off? If results are not where they should be, that could be an indication that adoption has not taken place or the solutions put into place are not working.

Getting Processes to Stick

The means to get people to adopt and change will happen through a series of campaign events that take place in a progression. The sequence of these events, their focus, and content can differ based on the type of stakeholder receiving the communications. The best approaches typically use repeated communications over a period of time that allow people to absorb change at a pace that they feel comfortable with. The picture below outlines an example of a sequence of campaign events for a new IT solution (process, new tool and new responsibilities) to be learned by IT support staff:

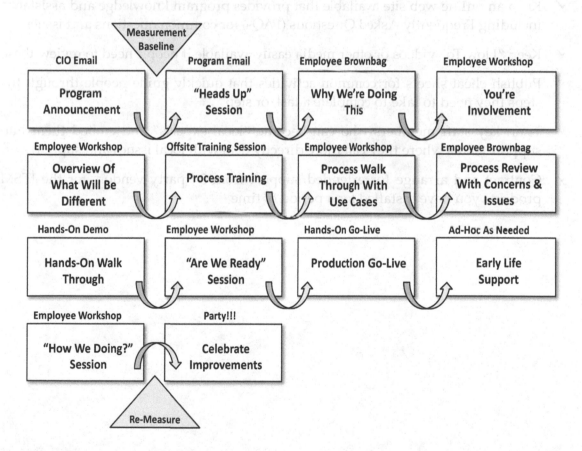

Note that in the above, a set of measurements has been taken at the start of the campaign and again at the end. Also note that there is a progression of communications that continually adds more and more to their skill sets, adoption and understanding. Also note that variety of event channels (workshops, demos, emails) were used to keep people engaged.

Early Life Support Strategy

Throughout the campaigns and even for a period of time post transition, an Early Life Support set of activities needs to be in place. It is not uncommon for people to have many questions, issues and some confusion over the changes that are now in place. A stepped up effort to support them is typically needed during this time. This can be wound down as people get more comfortable with the new solutions.

Some examples of early life support activities can include:

✓ Having Organizational Change Team members maintain a physical presence in staff work areas to mentor and support staff as needed

✓ Establish a hotline that is easily accessible where people can go for quick answers to their questions

✓ Provide additional training and review sessions post transition that people can attend if they feel they need more help

✓ Keep an online web site available that provides program knowledge and assistance including Frequently Asked Questions (FAQs) for common questions and issues

✓ Keep "How-To" videos or other media easily available if people need to review these

✓ Publish cheat sheets for common activities that quickly guide people through the steps they need to take to complete a task or step

✓ Train key staff members who can become "local experts" and embed them into support teams where they can more directly deal with local issues and support

✓ Contract and arrange for extended support from 3rd party vendors for the ITSM products you have installed for a period of time

Communication Tools and Techniques

This chapter presents a number communication tools and techniques. Additional ones can also be found on ITSMLIb as well as generally throughout the internet. Some ideas and concepts are presented here, but feel free to also use your own creativity. Anything is game!

Kaizen

Best Use: Embarking on a low-risk steady program of service improvement.

Kaizen is the Japanese term for a program of small improvements that individually provide small savings and efficiencies but in total amount to significant savings and efficiencies. Observe activities as IT support staff undertake them. Are there small manual tasks that are constantly being done over and over that frustrate staff? For example, many times staff are cutting and pasting information from one tool to another. Could something like this be automated without much effort? Will planned tool changes make things better?

Consider establishing a small program to first identify opportunities and then address them within the context of the new ITSM solutions you are going forward with. Utilize the Continual Service Improvement Register to inventory and track these small efforts. Execute a Kaizen Program as a logical team with the support staff involved. This can go a long way towards staff acceptance of new processes and tools. They can see small living examples of how things might get better and feel they had a hand in crafting those solutions. Moreover, they actually get tangible benefits from this that can be touted later to others.

Brainstorming

Best Use: Quickly generating ideas and consensus on how to solve a problem or issue.

This creative problem solving technique provides an efficient way to generate large numbers of possible ideas and gaining group consensus on which are the most important. The ideas themselves can be done with Post-It notes or electronic meeting software and applications that provide similar functionality (Especially great if your support teams are in multiple locations or global settings).

This technique works as follows:

1. Finalize the topic, mission and goals of the brainstorming session

2. Each person is given 5-15 minutes to write down as many ideas as they can, one idea per Post-It (or electronic posting)

3. In round-robin fashion, post each idea one at a time in front of the group, reading it to the rest of the group to make sure it is understood

4. When all ideas have been posted, collectively to look at them and begin grouping them into common or natural themes

5. For each group of ideas, title the group with a word or two the characterizes the entire group (e.g. "Skills", "Hardware Upgrade", "Software Fixes", or "Communications"

6. Summarize and review the groups with all participants for completeness

7. Have participants vote on the groups – an effective means to give each participant 10 votes and have them "spend" their votes on any of the groups (they can "spend" as many votes as they want on any one group)

8. Summarize and discuss the priority of the idea groups with participants to gain consensus

Brainstorming can be applied to many situations in order to get ideas and group consensus on a direction or decision. Some examples of its use can include:

Issues-Finding

Looking at the "mess" or interrelated issues, challenges, problems, and opportunities to find an area on which to focus. Idea groups should list significant challenges and opportunities, then isolate the main ones.

Fact-Finding

Exploring symptoms, unknowns, issues, challenges, missing or needed information to expand understanding of the mess. Include what you don't know as well as what you know in these sessions.

Problem-Finding

Discovering an underlying cause or issue behind a series of events, incidents or outages. List potential causes and then identify the best wording of the central problem.

Solution-Finding

Converging on a subset of ideas and then refining them into potentially useful solutions. List each solution and then agree the best one with the participants.

Action-Finding

Generating and refining potential action steps to implement a solutions. List action steps. Once grouped and agreed, assign names, resources and target dates to them.

Change Impact Modeling

Best Use: Modeling the impact of a coming change on stakeholders identifying the risks involved and actions to mitigate those risks.

With this technique, inventory what is changing in the IT organization and identify the risks and impact. Then identify the approaches to be taken for mitigating the risks. Summarize these on a spreadsheet. This might look as follows for a new Change Management initiative:

Change	Severity	Groups Impacted	Risks	Mitigation
Conversion from familiar custom-built app to unfamiliar tool	High	All IT Support Teams Change Management Team	Tool not usable Cannot get off old tool licenses, support staff may not enter their changes	Ensure all stakeholders are well trained in the new tool
CAB has different membership	Medium	CAB Members, Service Owners	Changes won't be properly authorized	Train new CAB members in their responsibilities
New CMDB for assessing impact of changes	Medium	CAB Members, IT Support Staff	Change impact won't be properly assessed	Train support staff in use of new CMDB and ensure CI relationships are in place

Here is some guidance for the columns in the above table:

Change

Anything that will be done differently that represents a significant change in how people work and operate. This column essentially answers the question about your solution: "What is it that is changing?"

Severity

While this is subjective, this information can be used to help prioritize your change management activities and focus your attention where it needs to be focused most. The final assessment of the severity of a change also depends on other factors in the chart such as risk and impact.

Groups Impacted

List the stakeholder types that will be impacted by the change. The value of this information is not only to identify who is impacted, but also to get a feel for the size of the stakeholder audience in question.

Risk

List the risks involved with the change. What can go wrong? What might happen if the change cannot take place?

Mitigation

In this column, list what needs to be done to mitigate the risks identified and ensure the change will take place. Keep this at a high level. Items in this column can form a quality assurance checklist for the communications program to ensure it has properly included all needed items within its scope.

Organizational Change Planning Model

Best Use: Assembling all activities into a cohesive visual planning model that can be communicated to executives and staff.

The model below presents one example of how you highlights key work streams and activities in a people, change and learning approach that can apply to an Organizational Change program. The model can be used as an overall framework from which you can populate work streams and tasks:

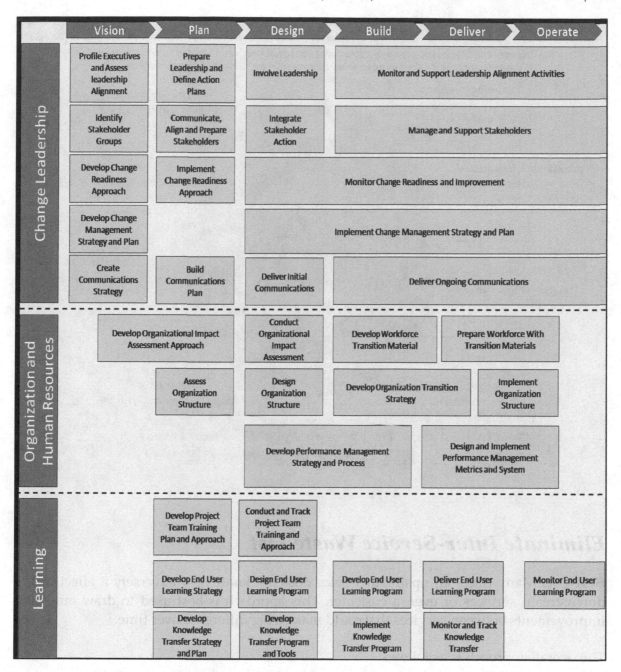

	Vision	Plan	Design	Build	Deliver	Operate
Change Leadership	Profile Executives and Assess leadership Alignment	Prepare Leadership and Define Action Plans	Involve Leadership	Monitor and Support Leadership Alignment Activities		
	Identify Stakeholder Groups	Communicate, Align and Prepare Stakeholders	Integrate Stakeholder Action	Manage and Support Stakeholders		
	Develop Change Readiness Approach	Implement Change Readiness Approach	Monitor Change Readiness and Improvement			
	Develop Change Management Strategy and Plan		Implement Change Management Strategy and Plan			
	Create Communications Strategy	Build Communications Plan	Deliver Initial Communications	Deliver Ongoing Communications		
Organization and Human Resources		Develop Organizational Impact Assessment Approach	Conduct Organizational Impact Assessment	Develop Workforce Transition Material	Prepare Workforce With Transition Materials	
		Assess Organization Structure	Design Organization Structure	Develop Organization Transition Strategy	Implement Organization Structure	
			Develop Performance Management Strategy and Process	Design and Implement Performance Management Metrics and System		
Learning		Develop Project Team Training Plan and Approach	Conduct and Track Project Team Training and Approach			
		Develop End User Learning Strategy	Design End User Learning Program	Develop End User Learning Program	Deliver End User Learning Program	Monitor End User Learning Program
		Develop Knowledge Transfer Strategy and Plan	Develop Knowledge Transfer Program and Tools	Implement Knowledge Transfer Program	Monitor and Track Knowledge Transfer	

Change Risk Wheel

Best Use: Provides a rapid means for identifying critical risks that need to be identified in the change management plan.

The chart below provides a means for identifying and communicating potential risks with a Change Management strategy or plan associated with an ITSM effort:

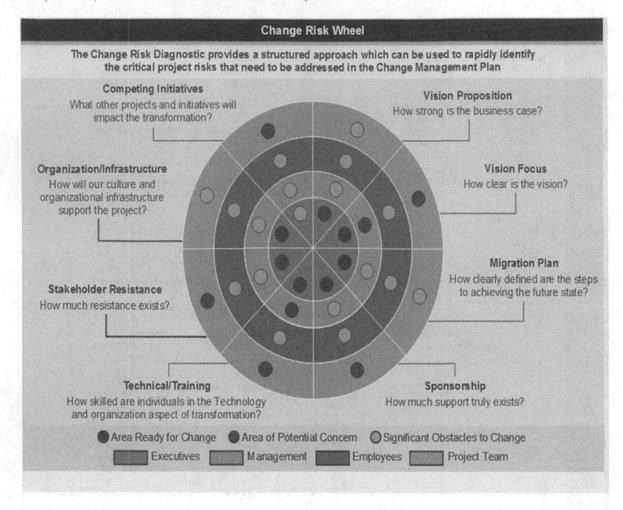

Eliminate Inter-Service Waste List

Best Use: Many times an upstream service does not know how adversely it effects other downstream services or even a customer. This approach is best used to draw out small improvements between services that could make a big difference over time.

The overall approach is as follows:

1. Assemble service owners and representatives for each service involved.

2. Two services that rely on each other (customer-supplier relationship in any form) agree or even make a pact to change practices or procedures to reduce work for the other area.

3. Each service owner describes the 5-7 most irritating things that need improvement (it should be mostly in the other service owner's control).

4. Estimate the frequency and the time spent to get an annual time spent on each item identified.

5. Each service owner shares their list of items with each other to understand how it impacts the other owner's service and then communicates it throughout. Don't be surprised if what is a minor mistake in one area creates significant effort to correct in another.

6. Both service owners then act to do something to reduce the occurrence of the wasteful activity or error and measure its results.

7. Once the improvements are working, each service owner devises a plan to sustain the improvement over time. This includes agreed upon periodic measurement to assure there is no slippage back to the old way of doing or not doing things.

With both owners working together to fix problems in their own service, corresponding efforts will be ensured. This will create a stronger bond between services as they see they are a part of the bigger picture of adding value to the ultimate customer.

Town Hall Kick Off Meetings

Best Use: Kicking off your change initiative with a major "Let's Start!" event. Somewhat like firing a starting gun in a race.

These meetings are gatherings of stakeholders to communicate the announcement of a new initiative or cover an issue of importance. They are somewhat ceremonial in nature. A location is identified where all stakeholders will meet and the meeting itself is conducted formally.

Most of the time, this technique might be used to announce a new ITSM program or a major change in direction. A typical agenda might look as follows:

- Introductions – "why you are here…"

- Summary of the change about to take place

- Benefits of the change

- Timing of key change events

- How this will impact each stakeholder

- Next steps or what they can expect to happen next

Optionally, you can include breakout sessions with these to discuss details unique to each stakeholder type. You may also add some time to address questions or concerns that people may have about the change they are about to undertake.

The takeaway from these meetings is that stakeholders fully understand the nature and business reasons for the change taking place. They see how things will evolve and how any concerns they have are being addressed.

It is absolutely imperative that senior executives participate in these meetings, even if only at the start. This displays the commitment and importance of what will be taking place. If they do not participate, it sends a negative message that will take the wind out of the entire effort.

Information Mapping

Best Use: Creating highly efficient documentation for policies, processes and procedures that is easy to absorb and gets right to the point without any extraneous information.

This is a neat technique for documenting processes or other instructions for stakeholders. The general concept is to avoid length discourses on information and scale it down to just the bare information that a particular stakeholder needs to get their jobs done. While much information about this technique is on the internet, here is a summary of some of the key points:

- Information should be communicated in small chunks and only relevant to what is needed to perform a job or task.

- Documentation consists of lots of white space and is formatted in easy to read fonts, tables and lists.

- There is a big focus on what and how tasks need to be done versus lengthy explanations on policy or historical discourse that stakeholders don't care about.

- Navigation is easy and straight forward. A stakeholder looking for a specific chunk of information should be able to find it easily without having to read through lengthy paragraphs or the entire document.

Proponents of this technique love to do before and after comparisons of documentation. Below is a small example:

BEFORE EXAMPLE: How To Submit An IT Change RFC

The Change Management policy was first enacted seven years ago to ensure that all application and infrastructure changes undergo management review and approval. The Change Control Meeting Process provides a single point of contact for review of upcoming submitted and planned changes and a forum for all affected groups to discuss the proposed changes. This provides all involved IT team members a better outline of the impact on both internal functions as well as external services facing the user community. Change control is open to all ITS staff and meets each Tuesday at 4:00 PM sharp in the room 6 facility. The change control group also discusses outcomes of changes from the previous week, change related information pertinent to IT technical groups (Change Freezes, Windows, Process and Procedures), and in general acts as a forum for all things under the auspices of Change Management. Before undertaking the change, first make sure you have a completed RFC with full authorization.

The Change Initiator is responsible for creating and finalizing the RFC form, consulting other stakeholders as required. As required, other documents may be referenced. The Change Initiator then submits the RFC and the Change Manager is notified. If an RFC is not approved during a review, the Change Owner may decide to Cancel, Re-Submit or Escalate the RFC. Escalation involves meeting with the next level of approver as discussed in Section 3.

This activity represents another exit scenario: During the process of planning, building, testing and deploying a change, the Change Initiator or Owner may decide that the change should be aborted or that the approach should be revisited. For instance, one of the entry or exit criteria may not be met (example: key business resources are not available). Once the next steps are determined, the RFC should be updated and the stakeholders should be informed. The process may then resume at the appropriate step (or the RFC may be cancelled).

AFTER EXAMPLE: Submitting An RFC

Submitting an RFC

Introduction	The following section describes the procedure for submitting an RFC when you are making a change to production.
Business Reason	Changes create the biggest risk for outages. If not properly recorded, we are exposed to outages and penalties.
Who Submits RFCs?	Any IT support staff, technician or employee that will be changing production hardware, software or network assets.

Continued on next page

Submitting an RFC *continued*

How To Submit An RFC	Use this procedure:	

Step	Action
01	Obtain an RFC form located on the Company XYZ SharePoint
02	Fill out all the fields on the RFC Form
03	Attach any supporting documents to the form (e.g. implementation plan, schematic, etc.)
04	Review the form with at least ONE of your peers
05	Hit the SUBMIT button on the form

How To Cancel an RFC	Use this procedure:	

Step	Action
01	Obtain your RFC form from the Company XYZ SharePoint
02	Set the change status field to CANCELLED
03	Hit the SUBMIT button on the form

Note the difference in style between the two examples. The second example strips out all the history of why change management is used and what it is. This is just noise that the reader doesn't really need. They just want to submit their RFC!

Also notice all the white space and how information is put into lists and tables versus a long narrative. The information is also chunked into bite-sized pieces with headers on the left so you can easily jump to the item of information you are looking for.

This technique has been around since the early 1980's. It was initially developed by Robert E. Horn. A search on the web under "Information Mapping" will yield a large trove of resources, examples and consulting firms that specialize in helping companies better document their policies and procedures. The book *Mapping Hypertext: The Analysis, Organization, and Display of Knowledge for the Next Generation of On-Line Text and Graphics* is a great resource for using this technique (ISBN 0-9625565-0-5).

Cheat Sheets

Best Use: Provide stakeholders with quick easily accessible information to accomplish common frequently performed tasks.

This is a simple but effective technique to support stakeholders in doing tasks in a different way. This consist of basic simple information about how to perform specific tasks hopefully avoiding calls to the Service Desk or interrupting others with questions for help.

These can be done in a variety of ways, but some general guidelines:

- Focus these directly on questions stakeholders may frequently ask.

- Document what needs to be done in a highly short, brief and to the point manner without unneeded information (see the Information Mapping described earlier)

- Make sure it fits on one page, or better yet, an index sized card

- Consider laminating it or putting in some format where stakeholders may post it near their workstation or desk

- Ensure that the information provided includes a contact number or web site URL where the stakeholders can go if they need more information.

Example CIO Global ITSM Message

The example below presents the text from an actual CIO of a global IT organization operating around the world. This text was from a global ITSM kickoff event.

"I wanted to give you an update on the ITSM Program – an Investment Committee top priority Program within Operations to improve service, enable transformation through the implementation of best practice Service Management processes and systems while allowing us to achieve reductions in operational costs. ITSM will enable a common way of providing Service Management globally.

ITSM is backed by the widely acclaimed ITIL (IT Infrastructure Library) framework. ITIL is a globally recognized industry standard code of best practice for IT Service Management, promoting business effectiveness in the use of information systems. The Program will deliver best practice based on the ITIL framework against key Service Management processes, distributed across two core disciplines, Service Support and Service Delivery. Linking into end to end service management processes across the business, these will be applied across our data centers worldwide, consolidate the number of systems used in these areas to enable a common workflow, and provide technical ownership for the deployed infrastructure.

In Q4, an assessment of Service Management capability was conducted at five Data Centers across the globe by our consultants. The study identified that compared to best practice, we lag behind the industry in terms of its capability across many areas of Service Management. The report recommended a number of improvements across Service Support and Delivery.

The results of the survey showed that while we have hard working, capable Operations staff, we are well behind best practice in our Service Management capability. This is because we have inconsistent processes and too many different systems and platforms. Using the proven principles of ITSM, we now have a significant opportunity to achieve a consistent and high quality approach to service management globally.

As ITSM moves into its next phase, we will focus on delivering quick wins in the most urgent and visible areas. Action Plans to this effect will be drawn up by the end of Q1 for implementation in Q2. In parallel, the project team will be working closely with business areas that have a key stake in the provision of end to end service management. This will enable us to develop and implement process improvements to ensure that the benefits realized early in the Program can be sustained through process and system improvement.

Many of you will be involved in an ITIL education Program starting in Q2 covering ITIL best practice and how it relates to our Operations. Any enquiries or bookings related to ITIL training should be channeled through the ITSM Program.

For more information on the Program, please visit the ITSM website."

ITSM Vision Statement Workshop

Best Use: Quickly crafting an ITSM program vision statement.

Many times, IT organizations spend inordinate amounts of time trying to craft vision statements. Here is an approach that could be used to quickly get to one in 2 hours or less.

The table below describes a suggested agenda for this kind of a workshop. Prior to this, make sure that the right stakeholders will participate. These should be those with leadership or influencer roles within the IT organization.

Agenda Item	Time	Activities
Introductions	15 minutes	• Facilitator introduces self • Each member says name, role in IT organization or company • Each member states their personal ITSM vision in a simple sentence
Purpose and Overview	5 Minutes	• Review purpose and outcome of session • Explain the session agenda and what will be done
Values Assessment	20 minutes	• Review Corporate Vision Statement if one exists • Each member gets 10 sticky dots • Values list is shown on the wall (See Values List below) • Members spend however many dots they want on any of the values in the list • Facilitator ranks these by which ones got the most dots • Prioritized list is reviewed with group
Vision Drivers	20 minutes	• Definition of a Vision Statement is reviewed with group (See Vision Statement Concepts and Examples below) • Driver questions are shown on the board or flipchart (See Vision Statement Discussion Questions below) • Facilitator reviews these one at a time with members • Key points are documented underneath each appropriate question

Agenda Item	Time	Activities
Brainstorm Vision	45 minutes	• Each member gets a yellow sticky pad • Members get 15 minutes to write down a single word or sentence that they feel should be included in the vision statement • Facilitator then reads these individually and makes sure concept is understood by group • Facilitator then places these on a whiteboard and initially groups them into broad categories based on similarity • Yellow pad stickies are then placed together in like statement structures • Statements are then modified and streamlined based on group discussion • Use dots techniques similar to before if group cannot agree among several sentence options
Summarize and Wrap-Up	10 Minutes	• Review vision statement against results of value assessment to confirm agreement • Review vision statement for alignment with corporate vision statement if one exists • Note any issues or concerns if these have arisen • Describe any next steps to be done

Values List

A table similar to the one shown below can be put on the wall of the meeting room. Make sure it is large enough to be seen by the whole group and can accommodate the placement of dots or checkmarks.

Place Dots Here	Value Statements
	Reduce Risk
	Increase Revenue
	Decrease Costs
	Avoid Costs
	Improve IT Service Quality
	Improve Productivity Of IT Staff
	Decrease Time To Market
	Improve Customer Service
	Provide Competitive Advantage
	Become Recognized Market Leader
	Other (describe here...)

Vision Statement Concepts and Examples

The following criteria can be applied to vision statements:

- Mutually agreed statement of "where we want to be"

- Based on looking at forward IT and business objectives

- Should describe the aim and purpose for IT

- Characteristics include:

    - Clarifies direction of the Service management program

    - Motivates people to take positive action

    - Coordinates the actions of people, process, technology and governance

    - Outlines the views of senior management

    - Cab be explained in less than 2 minutes to each stakeholder

    - Not specific to products or a way of operating

Below are some example statements:

- Support COMPANY NAME business goals of transforming itself into a leading information company by providing the highest quality IT service infrastructure that delivers the most reliable, accurate and secure (BUSINESS SERVICE OR PRODUCT) services in the world.

- We will set the standard for how to operate a global IT infrastructure that is admired around the world – most importantly - by our customers as a reason to buy COMPANY NAME services.

- We will be the preferred partner for IT services to the business.

Vision Statement Discussion Questions

The following questions may be used to help workshop members think about possible vision statements:

- Who are our customers?

- How should we appear to our customers?

- How should we appear to our stakeholders?

- What must we excel at with our customers and stakeholders?

- How will we sustain our ability to change and improve?

Other Sources

Although not an exhaustive list, these are additional sources you may wish to investigate that provide knowledge, expertise and approaches around Organizational Change:

- ADKAR® Model For Individual Change is a widely used approach for controlling and managing individual change (www.prosci.com)

- McKinsey 7S Framework is a widely used model used to assess and monitor organizational change efforts (en.wikipedia.org/wiki/McKinsey_7S-Framework)

- Balanced Diversity – A Portfolio Approach to Organizational Change by Karen Ferris – TSO (ISBN 978-0-11-7080608-7)

- John Kotter has a number of great books on Organizational Change. Most notable are:

  - Leading Change - Harvard Business School Press (ISBN 978-0-87584-747-4)

  - The Heart of Change: Real-Life Stories of How People Change Their Organizations - Harvard Business School Press (ISBN 978-1-57-851254-6)

  - Our Iceberg Is Melting: Changing and Succeeding under Any Conditions - St. Martin's Press (ISBN 978-0-31-236198-3)

- Managing Change by Robert Heller – DK Publishing (ISBN 0-7894-2897-0)

- Mel Silberman's The Consultant's Big Book of Organization Development Tools : 50 Reproducible Intervention Tools to Help Solve Your Clients' Problems - McGraw-Hill Education (ISBN 978-0-0714-0883-7)

- Skills Framework for the Information Age (SFIA) at http://www.sfia-online.org is a wonderful site for all kinds of organizational guidance and has suggested job descriptions for just about any IT role or position

- Change Management Model - Understanding the Three Stages of Change – Kurt Lewin at http://www.change-management-coach.com/kurt_lewin.html

ITSM Operating Roles

Overview

This section describes key operating roles needed to operate an IT service organization. The roles discussed here include the following:

- Steering Group Member

- ITSM Program Manager

- Project Manager

- Process Owner

- Core Team Member

- Extended Team Stakeholder

- Advisor Team Stakeholder

- Subject Matter Expert (SME)

- Process Architect

- Tool Architect

- Tool Developer

- Organizational Change Leader

- Organizational Change Analyst

- Facilitator

- Trainer

- Training Coordinator

- Technical Writer

- Coalition Team Leader

- Coalition Representative

- Administrative Analyst

Each role is described along with a suggested set of skill traits needed to be successful in each role.

Skill Levels

The following skill traits were identified as fairly typical of those needed to execute many of the key activities identified:

- Customer Relationship

- Negotiation

- Project Management

- Technical Architecture?

- Process Architecture

- Business Skills

- Communications

- Leadership

- Writing

- Teaching/Coaching

- ITIL/ITSM

- Administrative

- Analytical

- Political/Social

- Planning

- Operational Expertise

- Problem Solving

To handle skill levels a simple skills rating system from 0 to 5 and related criteria has been recommended. Again, you may wish to alter this based on the general practices of your organization and your specific program needs. The suggested skill levels and criteria are as follows:

Skill Level	Criteria	Description
0	Skill not necessary or needed	Cannot perform this skill and has little or no knowledge of where the skill might be needed or how it could be applied
1	Awareness of what has to be done	Understands importance of the skill and where it could be applied, but depends on others to use skill
2	Basic level but needs supervision	Has basic capabilities but still requires supervision, assistance or guidance
3	Can perform many tasks without supervision	Can perform this skill with little or no assistance
4	Can work independently	Can perform this skill and anticipates situations where the skill is needed
5	Can lead others	Possesses exceptional capability to perform skill, regularly coaches others on skill

Role Descriptions

The following section lists all the roles identified previously along with their definition, key activities, skills and skill levels. Note that these are recommendations only. Feel free to add, change or delete as you see fit. The purpose here is to give you a jump start on pulling your organization together.

Steering Group Member

This role sets project direction, makes key decisions and provides final approval of Program deliverables. Key activities for this role include:

- Champions process solutions across the enterprise

- Conducts periodic meetings to monitor Program progress and issues

- Provides final review and approval of program deliverables

- Coordinates approvals from business units as necessary

- Identifies and appoints key Program team members

- Coordinates major program decisions that have been escalated to the Steering Group on a timely basis to meet program objectives

Key skills for this role are:

Skill Trait	Skill Level
Customer Relationship	5
Negotiation	5
Project Management	2
Technical Architecture	1
Process Architecture	1
Business Skills	5
Communications	5
Leadership	5
Writing	2
Teaching/Coaching	2
ITIL/ITSM	1
Administrative	2
Analytical	4
Political/Social	5
Planning	4
Operational Expertise	2

ITSM Program Manager

This role ensures Program Work Products are delivered on a correct and timely basis and ensures the objectives of the Implementation Program are met. This role has oversight over all implementation activities and manages and monitors the overall program effort. Key activities for this role include:

- Responsible for the overall project objectives.

- Provides direction to the project teams for work products due as well as the overall status of the project.

- Assigns Initial Win and Process Implementation projects to Project Managers.

- Provides status of work in progress and/or issues to the Executive Steering Committee

- Develops project work plan, schedule and staffing requirements.

- Communicates as required to executive management.

- Conducts weekly change, issues and status meetings to track progress and risks.

- Ensures that outstanding project management, process implementation and design requirements and/or issues are being addressed.

- Communicates activities and status of the overall Program throughout its lifecycle.

- Schedules workshops and meetings as required.

- Provides overall leadership and management of the overall Program.

- Coordinates activities of Project Managers and Project Office staff.

Key skills for this role are:

Skill Trait	Skill Level
Customer Relationship	5
Negotiation	5
Project Management	5
Technical Architecture	2
Process Architecture	3
Business Skills	5
Communications	5
Leadership	5
Writing	2
Teaching/Coaching	2
ITIL/ITSM	2
Administrative	3
Analytical	4
Political/Social	5
Planning	5
Operational Expertise	3

Project Manager

This role provides project management oversight and expertise to assist Core Teams in accomplishing their objectives. Project Managers work within the Program Office and are assigned to Initial Win and Process Implementation projects. Key activities for this role include:

- Responsible for assigned project objectives.

- Provides direction to the project teams for work products due as well as the overall status of the projects assigned.

- Co-ordinates activities with other project managers when necessary.

- Provides status of work in progress and/or issues to the Program Manager

- Develops project work plans, schedules and staffing requirements for projects assigned.

- Communicates as required to executive management or Program Office staff.

- Conducts weekly change, issues and status meetings to track progress and risks with Core Teams assigned to.

- Ensures that outstanding project management, process implementation and design requirements and/or issues are being addressed for projects assigned.

- Escalates cross project issues or key management issues to the Program Manager.

- Communicates activities and status of each project assigned throughout its lifecycle.

- Schedules workshops and meetings as required.

- Provides overall leadership and management for the projects assigned.

Key skills for this role are:

Skill Trait	Skill Level
Customer Relationship	5
Negotiation	5
Project Management	5
Technical Architecture	2
Process Architecture	3
Business Skills	4
Communications	5
Leadership	5
Writing	2
Teaching/Coaching	3
ITIL/ITSM	2
Administrative	3
Analytical	4
Political/Social	5
Planning	5
Operational Expertise	3

Service Owner

Effectively the "CEO" of a service. This role owns one or more services end-to-end. It is accountable for the overall quality of a service, identifying improvements, measuring and meeting service goals and objectives.

Key activities for this role include:

- Single point of accountability for a service

- Measures the quality of services owned

- Ensures service is meeting business goals and objectives

- Proactively taking actions to improve services owned when necessary

- Establishing agreements with other IT service owners and third party suppliers to ensure effective service delivery is maintained

- Represents the service in change advisory board (CAB) meetings

- Ensuring that the service entry in the service catalogue is accurate and is maintained

- Participating in negotiating service level agreements (SLAs)

Key skills for this role are:

Skill Trait	Skill Level
Customer Relationship	4
Negotiation	5
Project Management	3
Technical Architecture	2
Process Architecture	5
Business Skills	3
Communications	3
Leadership	5
Writing	3
Teaching/Coaching	4
ITIL/ITSM	5
Administrative	2
Analytical	4
Political/Social	3
Planning	4

Skill Trait	Skill Level
Operational Expertise	4

Service Manager

This role ensures a service is operating and executing successfully on a day-to-day basis.

Key activities for this role include:

- Day-to-day operational responsibility for a Service

- First point of escalation for incidents beyond the Service desk

- Monitors work queues for support tasks

- Monitors service delivery activities for successful operation

- Coordinates activities to resolve Service incidents and requests across ALL support teams

- Coordinates major incident handling and response with problem management team

- Reviews changes for submission to change management process

- Works with problem management to proactively identify and address day to day service issues

- Proactively manages knowledge that will result in first call resolution

Key skills for this role are:

Skill Trait	Skill Level
Customer Relationship	3
Negotiation	3
Project Management	3
Technical Architecture	3
Process Architecture	3
Business Skills	3
Communications	4
Leadership	4
Writing	3
Teaching/Coaching	4
ITIL/ITSM	4
Administrative	3
Analytical	4

Skill Trait	Skill Level
Political/Social	4
Planning	4
Operational Expertise	4

Process Owner

This role ensures executive support of assigned ITSM processes, coordinates the various functions and work activities at all levels of a process, provides the authority or ability to make changes in the process as required, and manages assigned processes end-to-end so as to ensure optimal overall performance. Process Owners work with one another ensuring that process changes and improvements benefit the whole rather than help a specific function at the expense of another.

Key activities for this role include:

- Responsible for the overall process objectives.

- Provides direction to the process Core Teams for work.

- Monitors process maturity and progress throughout the implementation effort.

- Co-ordinates design decisions and activities with other Process Owners.

- Assists in development of project work plans, schedules and staffing requirements from a process perspective.

- Communicates as required to executive management and Program Office.

- Ensures that process implementation and design requirements are adequately identified and that process solution issues are being addressed.

- Identifies process and solution requirements to Technical Architecture Team.

- Identifies needed workshops and meetings as required to design and build process solutions.

- Coaches and teaches others about process concepts and solutions.

- Participates at communication events organized by the Organization Change Team.

- Provides overall leadership and management from a process perspective.

Key skills for this role are:

Skill Trait	Skill Level
Customer Relationship	4
Negotiation	5

Skill Trait	Skill Level
Project Management	3
Technical Architecture	2
Process Architecture	5
Business Skills	3
Communications	3
Leadership	5
Writing	3
Teaching/Coaching	4
ITIL/ITSM	5
Administrative	2
Analytical	4
Political/Social	3
Planning	4
Operational Expertise	4

Core Team Member

This role provides heads down implementation of ITSM solutions. It communicates with users of the process and with tool developers to implement the process. It also communicates with the Process Owner to receive direction and to provide feedback on how well the process is being implemented. This role also communicates with the Tool Architects for interfacing processes and tools to ensure integration of the process with other processes.

Key activities for this role include:

- Assists in development of project work plans, schedules and staffing requirements

- Communicates with users of the process as to what is expected of them

- Assesses the current state of readiness and effort required to implement the processes, tools and organization.

- Coaches the users of the process on tools and procedures.

- Communicates with the Process Owner on process design, status and issues.

- Manages resources during detailed solution design and implementation.

- Ensures that process documentation is developed and maintained.

- Participates at communication events organized by the Organization Change Team.

- Manages changes to tools and organization to support the process as required.

- Identifies additional resources as required to complete tasks such as writing procedures, developing job descriptions, producing analytical statistics or developing education material.

- Ensures interfaces to other processes are working well.

Key skills for this role are:

Skill Trait	Skill Level
Customer Relationship	3
Negotiation	3
Project Management	3
Technical Architecture	1
Process Architecture	4
Business Skills	3
Communications	3
Leadership	4
Writing	3
Teaching/Coaching	4
ITIL/ITSM	4
Administrative	2
Analytical	4
Political/Social	3
Planning	3
Operational Expertise	2

Extended Team Stakeholder

This role actively participates in the development of ITSM work products and solutions. Responsible for representing the business or IT unit interests in the solutions being developed, managing communications between the Core Team and their department and obtaining departmental approval of process solutions being developed.

Key activities for this role include:

- Actively assists in the review and development of ITSM Work Products.

- Provides input on solutions being developed to the Core Team.

- Coordinates decisions and feedback to the Core Team for key implementation and design decisions on behalf of the business units represented.

- Coordinates collection of solution requirements on behalf of the business units represented and feeds these to the Core Team.

- Obtains consensus and agreement on the solutions being developed from the business units represented.

Key skills for this role are:

Skill Trait	Skill Level
Customer Relationship	3
Negotiation	4
Project Management	1
Technical Architecture	0
Process Architecture	1
Business Skills	4
Communications	3
Leadership	3
Writing	1
Teaching/Coaching	1
ITIL/ITSM	1
Administrative	1
Analytical	3
Political/Social	4
Planning	3
Operational Expertise	1

Advisor Team Stakeholder

This role provides input and/or key decisions and recommendations to the Core Team on the solutions being implemented. It may also involve assigning others within the business unit represented to serve as Extended or Advisor Team members.

Key activities for this role include:

- Reviews output of the implementation effort and provides feedback to Core Teams.

- Provides key decisions and approvals on a timely basis to meet implementation project needs.

- Assigns other department personnel to serve as additional Advisor and Extended team members as needed.

- Works in conjunction with other Advisor or Extended Team members within the department as needed.

Key skills for this role are:

Skill Trait	Skill Level
Customer Relationship	3
Negotiation	4
Project Management	1
Technical Architecture	0
Process Architecture	1
Business Skills	4
Communications	3
Leadership	4
Writing	1
Teaching/Coaching	1
ITIL/ITSM	1
Administrative	1
Analytical	4
Political/Social	4
Planning	2
Operational Expertise	1

Subject Matter Expert (SME)

This role provides expertise in technical, business, operational and/or managerial aspects for the design and implementation. Participation in the implementation is as required. This role may also provide specialized expertise in the design and implementation of process solutions as needed.

Key activities for this role include:

- Provides technical, operational, business and/or managerial subject matter expertise.

- Provides input into the design of the procedures, tools or organization as required.

- Assists in the development of ITSM solutions by providing specialized expertise as required.

- Supports the development and execution of test scenarios designed to validate the functionality of the design.

- Validates the Design and Implementation Team designs for processes, tools and organization and any recommendations.

- Provides consultative and facilitation support to the Implementation Project Teams.

- Assists in creation of project work plans and implementation strategies.

- Provides Intellectual Capital as required during the Implementation Project.

- Coaches team members in specialized skill sets if required.

Key skills for this role are:

Skill Trait	Skill Level
Customer Relationship	2
Negotiation	2
Project Management	1
Technical Architecture	See Note
Process Architecture	See Note
Business Skills	See Note
Communications	See Note
Leadership	3
Writing	1
Teaching/Coaching	3
ITIL/ITSM	See Note
Administrative	1
Analytical	4
Political/Social	2
Planning	2
Operational Expertise	See Note

(Note: Level 5 skills should exist in subject matter area which may be any one of the items listed with "See Note")

Process Architect

This role establishes the overall strategic process architecture and ensures a well-architected solution from a process perspective. The role provides consultative help around process modeling, design, build, implementation and rollout. One primary benefit and focus of this role is to coordinate common activities between the Process Core Teams to ensure maximum efficiency.

Key activities for this role include:

- Coordinates all process activities across process Core Teams to ensure that ITSM solutions are integrated from a process perspective.

- Provides process expertise and input into the design, build and implementation of the ITSM solutions as required.

- Reviews process solutions under design for efficiencies in performance and resources to limit non-value labor and waste.

- May participate and lead in process modeling activities if used to design the ITSM solution.

- Supports the development and execution of test scenarios designed to validate the functionality of processes being designed and built.

- Supports governance activities from a process perspective.

- Coaches team members in specialized process skill sets if required.

Key skills for this role are:

Skill Trait	Skill Level
Customer Relationship	2
Negotiation	4
Project Management	1
Technical Architecture	1
Process Architecture	5
Business Skills	2
Communications	2
Leadership	4
Writing	3
Teaching/Coaching	3
ITIL/ITSM	5
Administrative	1
Analytical	4
Political/Social	1
Planning	2
Operational Expertise	3

Tool Architect

This role establishes the overall strategic tools architecture and ensures a well-architected set of technical solutions to support ITSM initiatives. The tool architect plays a dual role. It identifies and implements appropriate technology solutions that support the goals of the implementation. It also identifies new and changing technology solutions emerging in the marketplace that could provide value to the ITSM solutions being developed. This role also coordinates common technology related activities between all project teams involved in the implementation effort.

Key activities for this role include:

- Ensures the tool architecture meets the strategic needs of the implementation effort.

- Coordinates technology product selections and tailoring.

- Supports cross team early launch planning from a technology perspective.

- Ensures maximum integration of tools.

- Coordinates technology product implementation activities.

- Coordinates technology customization and integration activities.

- Coordinates Technical resources to optimize use of technology solutions.

- Identifies ongoing support and maintenance for technologies chosen.

- Communicates chosen tool architectures and solutions to program teams.

- Interfaces to technology vendors as needed.

- Provides information on new and changing technologies to implementation teams.

Key skills for this role are:

Skill Trait	Skill Level
Customer Relationship	2
Negotiation	4
Project Management	3
Technical Architecture	5
Process Architecture	1
Business Skills	2
Communications	2
Leadership	4
Writing	2

Skill Trait	Skill Level
Teaching/Coaching	1
ITIL/ITSM	2
Administrative	1
Analytical	4
Political/Social	1
Planning	3
Operational Expertise	5

Tool Developer

This role implements and customizes the technologies chosen to support the ITSM solutions being implemented. Activities also include provision of technical support during the implementation effort and assisting training activities by setting up training environments as necessary.

Key activities for this role include:

- Understands the processes, tool requirements and data requirements for the technologies being implemented.

- Provides input to the detailed design for the processes based on technologies planned for implementation.

- Customizes implemented technologies based on the detailed design.

- Tests technologies for expected operation.

- Assists in developing procedures to install technologies.

- Assists training activities by installing and customizing the education technical environment.

- Implements technologies into test and production environments.

- Resolves problems with technologies.

- Provides technical support for technologies throughout the implementation effort.

Key skills for this role are:

Skill Trait	Skill Level
Customer Relationship	1
Negotiation	1
Project Management	2
Technical Architecture	3
Process Architecture	1
Business Skills	1
Communications	1
Leadership	2
Writing	3
Teaching/Coaching	1
ITIL/ITSM	1
Administrative	1
Analytical	4
Political/Social	1
Planning	2
Operational Expertise	3

Organizational Change Leader

This role develops and leads the organizational change effort to alter business culture and behaviors towards alignment with the solutions being implemented. It also serves to build the change communication strategy and develop the communications plan. It monitors and oversees all ITSM stakeholders and carefully crafts and controls all key messages about the ITSM effort, its progress, stated vision and goals.

Key activities for this role include:

- Performs Stakeholder Management activities to identify Stakeholder concerns and issues with solutions being developed.

- Monitors stakeholder acceptance/rejection of solutions being developed.

- Crafts and controls key communications and messages about the implementation effort.

- Identifies opportunities to win acceptance of solutions being developed by those who are impacted.

- Identifies channels for communications and builds the overall communications plan.

- Develops a Resistance Management Plan to provide strategies for dealing with rejection or resistance to solutions being developed.

- Ensures appropriate levels of the organization are involved and demonstrating active commitment and leadership to the solutions being developed.

- Coaching senior management and other key personnel to help them "walk the talk" and demonstrate commitment to the ITSM solution.

Key skills for this role are:

Skill Trait	Skill Level
Customer Relationship	5
Negotiation	4
Project Management	3
Technical Architecture	1
Process Architecture	1
Business Skills	3
Communications	5
Leadership	5
Writing	3
Teaching/Coaching	5
ITIL/ITSM	1
Administrative	1
Analytical	3
Political/Social	5
Planning	3
Operational Expertise	1

Organizational Change Analyst

This role supports the Organizational Change Leader with a variety of administrative and organizational change development tasks as needed to meet the goals of the ITSM implementation effort.

Key activities for this role include:

- Maintains stakeholder documentation as needed.

- Develops training and presentation materials.

- Schedules training for ITSM team members as needed.

- Prepares artifacts related to the ITSM communications strategy such as newsletters, program giveaways and communication reports.

- Maintains e-mail distribution lists for stakeholders and ITSM implementation Program personnel.

- Schedules key meetings with stakeholder teams and steering group members.

- Takes and publishes notes at key program meetings and workshops that involve stakeholders.

- Other administrative tasks as needed to support the implementation effort.

Key skills for this role are:

Skill Trait	Skill Level
Customer Relationship	3
Negotiation	2
Project Management	1
Technical Architecture	1
Process Architecture	1
Business Skills	2
Communications	3
Leadership	2
Writing	2
Teaching/Coaching	3
ITIL/ITSM	1
Administrative	3
Analytical	1
Political/Social	2
Planning	1
Operational Expertise	1

Facilitator

This role leads and conducts working sessions and meetings in a neutral fashion to ensure that the goals of those sessions and meetings are met.

Key activities for this role include:

- Leading meetings and working sessions in a neutral manner to ensure goals and outcomes of those sessions are being met.

- Developing session detail agendas and agrees these with those involved.

- Developing discussion strategies and methods to ensure all participants are involved and to obtain consensus on key decisions in an efficient manner.

- Monitoring sessions to make sure all sides of discussed issues are being considered and that session groups do not "go with the flow" unless truly in agreement.

- Identifying needed materials and supplies for meetings.

- Providing feedback to Organization Change Team on participant acceptance of meeting issues and activities based on observation during meetings.

Key skills for this role are:

Skill Trait	Skill Level
Customer Relationship	5
Negotiation	4
Project Management	1
Technical Architecture	1
Process Architecture	1
Business Skills	1
Communications	5
Leadership	5
Writing	4
Teaching/Coaching	4
ITIL/ITSM	1
Administrative	1
Analytical	1
Political/Social	5
Planning	3
Operational Expertise	1

Trainer

This role provides training for process procedures and use of processes and tools. It identifies training needs and requirements. It builds the needed curriculum paths for each ITSM implementation team member and stakeholder. It leads and conducts training sessions. It also leads in the development of training materials.

Key activities for this role include:

- Identify needed training and curriculum paths for ITSM implementation team members and business stakeholders.

- Training users on processes, use of tools and procedures.

- Leading development of training material as needed.

- Leading and conducting training sessions

- Designing and building training curriculum for implementation personnel and business units impacted by ITSM services

- Identifying ITSM certification needs and requirements

- Coordinating Subject Matter Experts to assist with training as needed.

- Aligning training curriculum and events with Organization Change activities and plans.

- Publicizing training events and activities.

Key skills for this role are:

Skill Trait	Skill Level
Customer Relationship	4
Negotiation	3
Project Management	3
Technical Architecture	1
Process Architecture	1
Business Skills	1
Communications	5
Leadership	4
Writing	4
Teaching/Coaching	5
ITIL/ITSM	1
Administrative	1
Analytical	3
Political/Social	3
Planning	3
Operational Expertise	1

Training Coordinator

This role provides administration support for training activities. It assists with preparation of training material, manages training schedules, training registration and attendance. It tracks attendance at training and monitors status of training for each implementation team member and business stakeholder.

Key activities for this role include:

- Preparing training material as needed.

- Administering ITSM certifications and tracking certification results.

- Tracking attendance at training sessions and training progress for implementation personnel and business stakeholders.

- Administering training calendar and schedules.

- Handling administrative tasks associated with vendor provided training.

- Registering personnel for training activities and events.

- Other administrative tasks as directed by Trainers.

Key skills for this role are:

Skill Trait	Skill Level
Customer Relationship	3
Negotiation	1
Project Management	2
Technical Architecture	1
Process Architecture	1
Business Skills	1
Communications	3
Leadership	2
Writing	2
Teaching/Coaching	2
ITIL/ITSM	1
Administrative	4
Analytical	3
Political/Social	2
Planning	2
Operational Expertise	1

Technical Writer

This role sets standards for how processes and procedures are to be documented. It documents process guides and work instructions in a manner that is easily understood by those executing the processes. It participates in the documentation of tool architectures and tool changes. It builds and publishes templates for presentations and key ITSM forms.

Key activities for this role include:

- Setting standards for how processes and procedures should be documented.

- Produces documentation for process guides and procedures.

- Provides consulting guidance on how to best present documented information so it is quickly and easily understood.

- Identifies improvements for existing documentation.

- Designs and builds templates for key presentations and process work products

- Designs and builds templates for forms used as part of the ITSM solution.

Key skills for this role are:

Skill Trait	Skill Level
Customer Relationship	3
Negotiation	1
Project Management	1
Technical Architecture	1
Process Architecture	1
Business Skills	2
Communications	4
Leadership	2
Writing	5
Teaching/Coaching	2
ITIL/ITSM	1
Administrative	1
Analytical	4
Political/Social	2
Planning	2
Operational Expertise	1

Coalition Team Leader

This role leads and organizes activities and meetings with Coalition Team members for larger ITSM implementation efforts with many IT service organizations and delivery centers. It serves to ensure that ITSM solutions being developed will be able to be implemented across all the organizations represented. It also plays a main part in rolling out ITSM solutions to those organizations. It organizes coalition teams and leads coalition team activities. It also collects and summarizes input from Coalition team members for ITSM implementation teams. This role serves as an Extended Team Stakeholder.

Key activities for this role include:

- Organizing and leading Coalition Team meetings.

- Identifying the appropriate Coalition Team membership needed to adequately represent all the organizations it was established for.

- Identifying and obtaining Coalition Team members.

- Collecting and summarizing Coalition Team input and feedback on ITSM design decisions and solutions.

- Leading rollout efforts on behalf of the organizations to receive and operate ITSM solutions once they are built.

- Assisting ITSM solution design efforts by summarizing key ITSM related solutions that may be operating currently at the organizations represented.

Key skills for this role are:

Skill Trait	Skill Level
Customer Relationship	4
Negotiation	5
Project Management	4
Technical Architecture	3
Process Architecture	3
Business Skills	4
Communications	3
Leadership	5
Writing	2
Teaching/Coaching	1
ITIL/ITSM	3
Administrative	1
Analytical	4

Skill Trait	Skill Level
Political/Social	3
Planning	5
Operational Expertise	5

Coalition Representative

This role provides a single point of contact into one or more IT organizations and service delivery centers. It represents the concerns and ideas of those organizations. It provides input and feedback to ITSM solution designs and plans based on feasibility within the current infrastructure, operations and culture with the organizations represented. It identifies ITSM related solutions that may be operating in some of the organizations represented that may be of help to those designing and building ITSM services. It also assists with rollout and implementation of ITSM agreed solutions at the organizations represented. This role works with and reports to the Coalition Team leader.

Key activities for this role include:

- Reviewing ITSM plans, designs and key decisions with IT and business staff at the organizations represented.

- Providing feedback, concerns and issues that are raised by represented organizations to the Coalition Team Leader on ITSM plans and designs.

- Identifying ITSM related solutions already in place and operating that may be of help to ITSM implementation teams.

- Assisting in the development of rollout plans unique to organizations represented.

- Rolling out ITSM agreed solutions to the organizations represented.

- Attending Coalition Team meetings.

Key skills for this role are:

Skill Trait	Skill Level
Customer Relationship	4
Negotiation	4
Project Management	3
Technical Architecture	3
Process Architecture	3
Business Skills	3
Communications	2

Skill Trait	Skill Level
Leadership	2
Writing	2
Teaching/Coaching	1
ITIL/ITSM	3
Administrative	1
Analytical	3
Political/Social	3
Planning	4
Operational Expertise	5

Administrative Analyst

This role performs administrative and clerical duties and activities as needed to support the Implementation Program. This role mainly resides within the Program Office but may also exist within other teams as needed.

Key activities for this role include:

- Gathering and collating Program status report information.

- Administering Program document repositories and web pages.

- Collecting labor hour/time reporting information from Program participants.

- Managing Program Email distribution lists.

- Managing and publishing the Program calendar.

- Setting up Program Meetings and schedules.

- Coordinating travel arrangements for Program participants.

- Other duties as directed by the Program Office or other authorized team members.

Key skills for this role are:

Skill Trait	Skill Level
Customer Relationship	1
Negotiation	1
Project Management	2
Technical Architecture	1
Process Architecture	1
Business Skills	1
Communications	2
Leadership	1
Writing	3
Teaching/Coaching	1
ITIL/ITSM	1
Administrative	4
Analytical	2
Political/Social	1
Planning	1
Operational Expertise	1

Chapter
15

Training Considerations

Overview

The infrastructure that underpins IT services will involve the use of a variety of tools, technologies and procedures. In order to operate and maintain an effective infrastructure, support personnel must be trained in the use of these technologies and procedures. An effort to develop, coordinate and conduct a training program is needed to ensure that support personnel fully understand how to operate and support the service infrastructure.

Key areas for which training needs to be developed might include the following:

- ✓ Training in a basic understanding of new solution processes and their functionality
- ✓ Training on how to use new technologies that support those solution processes and functions
- ✓ Training to develop new skills needed to administer, support and maintain the toolsets and processes that underpin the new solution.

Training activities should be considered as communication events. Training paths and delivery take the form of campaign events. It is yet one more kind of communication that takes place to get others to adopt and use a new solution.

Key tasks specific to training include the following:

- ✓ Development of solution related training materials
- ✓ Coordination with other infrastructure projects as necessary to incorporate other training materials that those projects may have developed
- ✓ Coordination with 3rd party vendor training programs and activities
- ✓ Scheduling of training sessions
- ✓ Execution of in-house training programs
- ✓ Monitoring of training programs and student progress
- ✓ Control of off-the-shelf vendor training materials.

Kinds of Training for ITSM Solutions

As you put together your training strategies and plans, different kinds of training need to take place with the various stakeholder groups depending on their responsibilities. Examples of training can include:

Solution Functional Training

Training designed to give IT support staff an understanding and appreciation of new processes and how to use new tools in their job.

Technical Training

This is training to keep IT technical staff who will support the ITSM tooling solutions. It includes skills to support ITSM technologies, support for IT support staff and ongoing operations of those tools.

Operations Training

Training needed to allow IT operations staff to adequately support day-to-day solution support and delivery activities. While some training may be technically related to processing operations, other training may be more specific to the solution to resolve common problems using the new solutions.

Service Desk Training

This is training in solution specific procedures and skills for Service Desk staff needed to service users and direct incidents, requests and problems to appropriate personnel. This includes training in problem analysis techniques as well as customer relations and phone handling etiquette.

User Training

This is training that may be necessary to help users of ITSM solutions who will be using the new solution. Training should also include procedures to obtain related support they will need such as how to contact the Service Desk for example.

Training Delivery Strategies

As part of your communications plan, the approaches to be used for training need to be considered. These can differ based on the types of stakeholders who will be attending training as well as the size of the stakeholder audience to be trained.

The following skills transfer strategy can be considered as an overall framework for organizing training events. It progresses from simple awareness to mastery of the skills being taught. The learning and adoption stages in this strategy are shown in the following table:

Learning Stage	Learning Status	Training Content
Awareness	"What does this look like?"	• Overview of new solution • How this impacts them • What will happen next • Questions, issues and concerns
Apprentice	"Show Me How It Works!"	• Key processes and workflows • How tools will support their tasks • Demos and hands-on case walk-throughs • Questions, issues and concerns
Mentored	"Let me try it myself!"	• Knowledge guides and cheat sheets • Case walk-throughs with live examples • Transition and cutover plans • Questions, issues and concerns
Supported	"I can do it myself!"	• Mock transition walk-throughs • Transition and cutover • Implementation early support • Issues tracking and follow-up

Training events can be organized around these to create sequences of training events that carry stakeholders through from first awareness to mastery of new skills. A little more on these stages:

Awareness Stage

This includes training and communications about what is coming up. Its purpose is to first introduce stakeholders to new concepts, what they can expect and how those concepts will be applied to their work place. Training here is at a high level and deals mostly with concepts.

Apprentice Stage

In this stage, details of new processes and tools are presented. The focus of training is on how these are applied to the workplace. Students can see how tools actually work, for example, and walk through screen flows and other information as it will be applied to their work. The training dynamic is that trainers guide students through the material and show them how things are to be done.

Mentored Stage

At this stage the roles are reversed. The student now takes over and the trainer steps back into more of a mentor/assistant type role. Students are given exercises or actual tasks to perform. The trainer merely provides advice and steps in only when the student goes too far astray. During this stage, the mentor validates whether skills have been transferred to the degree that stakeholders are ready to use new skills and ways of working in the production environment.

Supported Stage

In this stage, students have successfully absorbed the new skills and are using them in their day-to-day work. Trainers are not necessarily needed other than to conduct post transition training sessions on an ongoing basis as needed. They may also assist in supporting students for a period of time under an Early Life Support program until such time this assistance is no longer needed.

A key design decision for training is to determine what delivery strategy will be used. Examples of different strategies and when to apply them include:

Traditional Classroom Training

Using a classroom setting to conduct face-to-face training and skills transfer with a live instructor.

Pros:

- Allows direct face-to-face contact

- Students can ask questions and get direct answers

Cons:

- Requires time away from work place

- May take significant time and effort if the audience to be trained is large (e.g. an audience of 400 may require 20 or more training sessions to be conducted)

Train-The-Trainer

Using a classroom setting to conduct an initial face-to-face training session with selected IT staff who will then go out and lead training face-to-face with others.

Pros:

- Best used for large or very large audiences

- Large audiences can more easily be trained in parallel shortening the overall time needed for training activities

- Allows face-to-face contact with students

- Students can ask questions and get direct answers

Cons:

- Requires extra time from IT staff who will be doing the training

- IT staff not only need to be trained in the new skills, but must also be trained in how to address student questions and issues

Super Users

Selecting a representative staff member who works within a business unit or office to act as the first line of support for that unit or office. The representative will become the "go-to" person to address questions, issues, concerns and even take on some support tasks as needed.

Pros:

- Provides a ready local site resource to assist stakeholders directly within their business units

- Leverages knowledge of new skills and tasks with day-to-day job tasks and ways of working (e.g. can address confusion with localized knowledge of things like "remember how we recorded special incidents in the old tool…now it is the same but the information can be entered like this…"

- Frees up calls and associated labor to the centralized Service Desk

- Can provide an accurate source to assess how well new tools and processes are actually working

- Can be leveraged to perform training and deployment tasks at the local level

Cons:

- Local site management needs to agree to provide resource commitment for this

- The risk of over-burdening the support resource needs to be managed carefully

- Communication of tooling and process changes and issues needs to be carefully managed to ensure super users are aware of changes that can impact them and the support they give.

3rd Party Trainer

Using a 3rd party supplier to provide training.

Pros:

- Provides an independent source for training free from any internal biases to students

- Allows direct face-to-face contact

- Students can ask questions and get direct answers

- Training materials typically come with the supplier shortening development time for training artifacts

Cons:

- Can be expensive, especially if training needs to take place on an ongoing basis to deal with staff changes post transformation

- Outside trainers will not be as familiar with internal issues and how things work, communicating in a manner that does not address those issues directly

- Care needs to be taken that adequate information is being provided versus sales presentations

Video Session

Creating training videos and making these easily available to IT staff.

Pros:

- Reduces need for physical labor to deliver training

- IT staff may attend training at the times that are most convenient for them

- Works well with large and very large audiences

Cons:

- Typically doesn't work as well for highly technical or complex process needs

- Costs for producing the videos need to be carefully considered

- Additional tools may needed internally to produce and record the videos

- Students have no ability to ask questions (although a support desk could be used to overcome this, it still disrupts their learning process)

- Tracking attendance (who has viewed the videos) can be a challenge unless carefully managed.

Use Case Walkthroughs

Presenting real life situations and then walking through the process steps and tools that will be used to deal with them.

Pros:

- Allows for live interaction with new processes and tools to address issues and situations that actually occur

- Can be good means for gauging whether knowledge transfer has actually taken place

- Students may uncover some situations or issues that the ITSM development teams have not considered

Cons:

- Creating examples of use cases that students actually experience can sometimes be a challenge

- Physical training settings must accommodate access to processes and tools for all students in attendance (although this could be done as a group activity, it won't be as effective if each individual student can get a hands-on training experience)

Lunch and Learns

Having a social event such as a special lunch or other gathering that is relaxing and social and use that to communicate new skills.

Pros:

- This approach is good for one-shot training on specific topics and subjects

- Greatly supports acceptance of new ways of working and can minimize resistance

- Students have an opportunity to learn in a more relaxed setting that is supportive

Cons:

- Social setting may not be conducive if there are issues or specific topics need to be presented with more detail

- Need to make sure that the social setting does not distract from knowledge transfer goals

Offsite Learning Events

This can be any of the other options except that these are held offsite away from the work place.

Pros:

- Creates a sharp focus for learning activities away from daily work distractions

- Concerns and issues can be raised in a more neutral setting and addressed by the whole group

Cons:

- Need to make sure that cell phones and other messaging distractions are not getting in the way of a successful learning event

- Can be expensive depending on location

- Typically must be planned for well in advance to make sure everyone can attend

Integration with Other Meeting Events

This involves taking advantage of other meetings that take place within the IT organization to conduct training. For example, there may be a general IT monthly meeting that all staff attend. In this case, a training session is added to the agenda for that meeting. This approach can also be used to provide training during CAB meetings, problem and incident meetings, Service Desk staff meetings or other events that typically take place.

Pros:

- Avoids additional scheduling efforts since those meetings are already in place

- Adds legitimacy to the ITSM Program (viewed as important since the meeting sponsor placed it on the agenda)

Cons:

- Time may be limited as the meetings also have other objectives

- Risk that some meeting agenda items may go overtime reducing the time available for the learning event

- For some meeting forums, it may be awkward to get into details or address specific issues

Self-Study Training

Providing training materials directly to students who then complete those materials.

Pros:

- This approach is good for filling in skills or increasing the current skills of existing personnel

- Students may take training at their convenience

- Reduces need for trainers

Cons:

- May incur longer development time to get materials ready for students

- Monitoring and tracking student completion of materials needs to be provided for

- Additional means need to be in place to distribute training and validate it has been completed

- The means for students to navigate and complete the training may incur additional development time

- Students may not have a good means for raising questions and getting additional help

Newsletters and Other Media

This approach is another informal way of disseminating information to staff. Examples might include newsletter type "how-to" articles or a "Didn't you know..." blurb on a company web site.

Pros:

- Works well for short blasts of information that people need to know

- Media may have an already built-in infrastructure for communicating to large audiences

Cons:

- May be constrained by deadline schedules for media publication (e.g. need the newsletter article by Friday or it won't be published)

- Not everyone actually reads all the newsletters or web site blurbs that are published

- More of a pull by students than a push – students must want to access media or they won't get the messages

Training Design Considerations

Training just doesn't happen by having an instructor show up. It needs to be carefully planned for and designed. Some key considerations related to training are as follows:

Training Responsibilities

Determination of who is responsible for what types of training. Areas of responsibility include the following:

- Preparation of training materials.

- Scheduling/Follow-up of training with personnel.

- Scheduling of instructors and facilities use.

- Monitoring personnel required training.

- Coordination with outside training vendors.

- Preparation of training materials.

- Updates to training materials.

- Training evaluation and maintenance.

Identification of Training Requirements

A training plan needs to be developed that outlines what skills are needed, how deep those skills need to be, and how they will be obtained. This plan may be directed towards practical skills or be further expanded to provide requirements for career paths and promotions.

Training Facilities

Facilities need to be provided for training activities. This includes items such as the following:

- Classroom space.

- Audio/Visual Materials.

- Study areas that are free from day-to-day disruptions.

- Duplicating, binding or other materials production facilities.

Training Costs

There may be various costs associated with training activities. Internal training may have costs for production of materials or personnel to conduct the training. Outside training may involve costs for registration, travel, meals and lodging.

Hands-On Training Needs

It may be desired to conduct training utilizing the actual system that is being learned. Preparation of this type of training if done on site will require the setup and maintenance of a training region or version of the solution being trained on. A separate training database may also be needed. These facilities should be isolated from other processing activities that may be occurring throughout the day. In addition, training databases should be recoverable and re-initialized for each class undergoing the training.

Modes of Training Delivery

Training may be delivered in a number of different forms. The appropriate form should be used as necessary to deliver training depending on skills needed and convenience for access to training. Some of these modes were presented in the Training Delivery Strategies section earlier.

Training Evaluation

Providing a means for feedback on training from trained personnel can be used as input to continually improve training delivery. Appropriate evaluation forms will need to be designed, given to personnel and analyzed to determine if valid improvements may be needed.

Establishing a Learning Environment

It is recommended that specialized facilities and locations be used to conduct training events. Within your communications plan, tasks need to be considered to ensure these facilities are put into place and that an environment conducive to training has been established. Some key considerations for this can include:

Demo or Training Room

Official location for providing hands-on tool training and walkthroughs. This should be setup in a classroom style with multiple concurrent access to tools by multiple users. The establishment of this can involve setting up multiple network connections and user devices with LOGIN access and standard student configurations.

Tool Training Environment

This involves establishing a tooling instance that can be used only for training purposes. Setup for this involves configuring the tooling instance and setting up training databases (tooling records that will be used in student exercises and walkthroughs). Planning for this should also include operational elements such as backing up training data and handling of outages that might occur during training times. The environment should be isolated from production, for example, generated email communications from the tool should be constrained to the training environment.

Training Materials

These are used for communicating skills. They can consist of training presentations, manuals, user guides or other media aids that will be used to communicate new skills. These will have to be designed, built and sometimes printed and bound to make them ready for training.

Support Facilities

This includes establishing a means for providing answers to student questions and "how-to" aids that students may need outside of the classroom. This can take the form of a student call hot-line, online web site or additional Service Desk support facility. Provision of any of these will require some up front planning, design and build activities.

Remote Laptop Instance

This is to support needs where training events may have to take place outside of the training facilities. It is not unusual to find that some training activities need to take place at remote locations or as part of a management session or support staff meeting. The instance needs to be provided with a network connection, LOGIN and standard training configuration. In addition, copies of training presentations, materials and aids should also be stored in case they are needed.

Online Conferencing

Facilities should be established to conduct online web conferences. This is a helpful vehicle not only for conducting some training events, but also to demonstrate how specific issues or situations might be handled.

Training Delivery Roles

Typical roles for design and delivery of training require staff with strong expertise in communications and how people learn. The key roles needed can include:

- Training Project Manager

- Training Support Analyst

- Technical Writer

- Trainer

A brief overview is below:

Training Project Manager

Provides oversight and project management over development and delivery of training. Key skills include:

- Solid skills in training development

- Development of training plans

- Training administration

- Knowledge of ITSM activities

- Negotiation skills to obtain management input and approvals

- General project management skills

Training Support Analyst

Responsible for development of training materials and training content. This also includes administration of training during delivery. Key skills include:

- Solid skills in training development

- Development of training plans

- Training administration

- Expertise in ITSM processes and/or tools

- Development of training materials.

Technical Writer

Responsible for writing, editing and production of training communications and artifacts. Key skills include:

- Solid skills with effective procedural writing
- Use of tools for production of training media
- Skilled with printing and production of training artifacts
- Solid written communication skills

Trainer

Responsible for delivery of training and knowledge transfer to those attending training. Key skills include:

- Good communications skills
- Good leadership skills to conduct sessions
- Skilled in the topic(s) being delivered
- Able to see how others are absorbing new skills and help them address any questions or issues they might have
- Respected by those attending training

Training Administrator

Responsible for coordinating and administrating training events. This includes activities such as scheduling attendees for events, tracking their participation, arranging for training space and coordinating setup and tear-down of training facilities. Key skills include:

- Good communications skills
- Coordination and management skills
- Detail oriented
- Solid administrative and reporting skills
- Ability to use administrative tools for scheduling, tracking and reporting

Training Plans

Each training event should have a separate plan and description. This should describe the learning objective(s) for the training and how that training will be conducted. The example below presents a sample plan for training support staff on a new incident management tool:

Incident Management Tool Training Plan

Learning Goal: To train support staff how to use the new Incident Management tool.

  Intent:

- Introduce support staff to the tool interface

- Provide an overview of the Incident Management process

- Train staff to a basic level of proficiency with the understanding that if they had to use the training received the next day they would have the skills to do so

Delivery Method:

- Instructor led, hands on training in a class room environment

- Provide exercises for creating and handling incident tickets to test staff understanding of the new tool and process

Learning Objectives:

- Introduce support staff to the tool interface and basic functionality

- Login as a support staff person to explain the incident process and tasks of the support team

- Demonstrate how to create and submit an incident ticket

- Demonstrate how to escalate an incident ticket to another support team

- Demonstrate how to review tool dashboard for incident tasks assigned to support staff

- Demonstrate how to close out an incident ticket

- Discuss issues and best practices as they relate to the incident management process

Measures of effectiveness:

- Support staff successfully completes the incident exercises

- Support staff can login and move around the new tool with little or no assistance

Training Audience and Size

Stakeholder Type	Audience Size
Senior Executives	5
Service Owners	20
IT Unit Leads	55
IT Operations Support Staff	22
Service Desk Agents	8
Technical Support Staff	41

About the Author

Randy A. Steinberg has extensive IT Service Management and operations experience gained from many clients around the world. He authored the ITIL 2011 Service Operation book published worldwide. Passionate about game changing management practices within the IT industry, Randy is a hands-on IT Service Management expert helping IT organizations transform their IT infrastructure management strategies and operational practices to meet today's IT challenges.

Randy has served in IT leadership roles across many large government, health, financial, manufacturing and consulting firms including a role as Global Head of IT Service Management for a worldwide media company with 176 operating centers around the globe. He implemented solutions for one company that went on to win a Malcolm Baldrige award for their IT service quality. He continually shares his expertise across the global IT community frequently speaking and consulting with many IT technology and business organizations to improve their service delivery and operations management practices.

Randy can be reached at RandyASteinberg@gmail.com.

Printed in the United States
By Bookmasters